WOW!
Resumes for
High Tech Jobs

Leslie Hamilton

McGraw-Hill

New York San Francisco Washington, D.C. Auckland Bogotá
Caracas Lisbon London Madrid Mexico City Milan
Montreal New Delhi San Juan Singapore
Sydney Tokyo Toronto

McGraw-Hill

A Division of The McGraw·Hill Companies

Copyright © 1999 by Leslie Hamilton. All rights reserved.
Printed in the United States of America. Except as permitted
under the United States Copyright Act of 1976, no part of this
publication may be reproduced or distributed in any form or by
any means, or stored in a database or retrieval system, without
the prior written permission of the publisher.

1 2 3 4 5 6 7 8 9 0 MAL/MAL 9 0 3 2 1 0 9 8

ISBN 0-07-026039-7

*The sponsoring editor for this book was Betsy Brown, the assistant
editior was Kurt Nelson, the editing supervisor was Fred Dahl, and the
production supervisor was Sherri Souffrance. It was set in Stone Serif
by Inkwell Publishing Services.*

Printed and bound by Malloy Lithographics, Inc.

McGraw-Hill books are available at special quantity discounts to
use as premiums and sales promotions, or for use in corporate
training sessions. For more information, please write to the
Director of Special Sales, McGraw-Hill, 11 West 19th Street, New
York, NY 10011. Or contact your local bookstore.

Dedication

To those who are ready, willing, and able to pester creatively.

Contents

Alphabetical Listing of Resumes

Acknowledgments

My gratitude goes out to the many people who helped me to turn this book into a reality. These include my editor Betsy Brown, who showed patience and support throughout the process; my friends Brandon Toropov and Glenn KnicKrehm, who offered important insights from the business world; Judith Burros, whose administrative and moral support was always extraordinary; Bert Holtje, who directly and indirectly, gave me hope to carry on; my husband Bob, without whom the book would never have come into existence; and my three daughters Meghan, Emma, Cassie, who always inspire their mother. Lastly James Voketaitis, of Resumes by James, provided invaluable help, insights, and assistance throughout the process. My deepest thanks go out to him.

Leslie Hamilton

About the Author

Leslie Hamilton (Boston, MA) is a writer and researcher who has written and contributed to numerous books in the areas of careers and personal finance.

Introduction

In Which You Discover the "Hmmmm" Factor

WOW! Resumes for High Tech Jobs is dedicated to the proposition that you can't win a job if nobody ever bothers to read your resume.

Your resume isn't—or at any rate shouldn't be—a dry, quasi-legal document (or for that matter an excuse to exaggerate your capabilities). It ought to be an advertisement that stops readers in their tracks and makes them long to find out more about you.

To the extent that your resume brings you closer to the ideal expressed in your written correspondence with employers, it will be successful. To the extent that it leaves people thinking, "Gee, this one really doesn't look much different from any of the others," it will fail.

This book features guidelines, examples, and checklists intended to help you make your resume stand out. Each chapter examines a particular resume challenge, highlights strategies you can use to make your resume stand out from the crowd, and offers sample resumes intended to help you fashion a resume that will help you win the numbers game, get the interview , (and eventually the job offer) you deserve.

When high tech employers review resumes and job queries, their thinking process sounds something like this:

KEY POINT:

Your resume's job is to win the battle of the declining attention span, and generate something akin to "Hmmmm" from a decision maker.

"What a huge stack of mail. Man, I don't know how I'm ever going to get through this one. Oh, well, better get started. Let's see. No. No. No. No. No. No. No. No. No. (Pause) Hmmmm......"

"Hmmmm" means an interview. "Hmmmm" means a chance to shine in person and win the offer you deserve.

In this book, you'll find strategies—and over 100 model resumes—that will help turn "No" into "Hmmmm . . ." and finally into "Yes."

Sometimes, simply following "the rules" is the worst way to go.

1
The High Tech Hiring Game

What Top Hiring Officials Want To See—and What They Don't

In creating, the only hard thing's to begin; a grass-blade's no easier to make than an oak.

JAMES RUSSELL LOWELL

Sometimes, simply following "the rules" is the worst way to go.

Take the task of finding a job within the field of high tech. While following what we perceive as established procedure may make the most sense when it comes to fulfilling the responsibilities of certain high tech-based jobs, it's not necessarily the best way to get the attention of a hiring official. As with most everyone else these days, the people who make decisions about whether to hire you are pretty busy. They need more than a factual recitation. They need to hear about realistic potential solutions to the difficult problems they face every day.

Peter F. Drucker recently wrote a piece in *The Wall Street Journal* (March 29, 1995) that touched on this topic. Drucker's remarks are worth reviewing closely here:

"Most resumes I get ... list the jobs the person has held," Drucker wrote. "A few then describe the person would like to get. Very few even mention what the person has done well and can do well. Even fewer state what a future employer can and should expect from that person. Very, very few, in other words, yet look upon themselves as a 'product' that must be marketed."

I believe what Drucker is calling for is a completely different type of resume than the one many of us are accus-

KEY POINT:

As a practical matter, you probably shouldn't expect a resume that obediently lists every position you've ever held and does little else to get superior results for you.

tomed to writing or reading—the kind that makes hiring officials break their routine and say "Hey, wait a minute!"—or better yet "WOW!"

The objective then is to formulate a WOW resume. This is a resume that takes responsibility not only for listing job titles, but also for supplying information about "what the person has done well and can do well." Some people will tell you that high tech is an inherently conservative field of endeavor, and that resumes for high tech professionals must, as a result, avoid the "marketing-centered" approaches that Drucker's remarks point to. I disagree.

In my view, aggressively highlighting possible solutions to the problems faced by decision makers on a daily basis is the only intelligent approach to resume development in today's employment market. What follows is some practical advice on developing such a WOW resume, and examples you can use as you craft your own.

Guidelines for High Tech Job Hunters

When it comes to high tech jobs, the good word for job hunters is "Hold the employers' feet to the fire!"

"Every company is running into the same problem," says Bruce Wideberg of New Boston Systems. "Companies are going all out to attract new employees and satisfy current employees. It has shifted from the Puritan ethic of 'live to work' to 'work to live' and the companies are beginning to understand this."

"The clients [companies] have a grasp of what they want, what they need, and what they are looking for people with experience, the right skills, and good personality. If they have these things, they [job seekers] are in a better position to negotiate," offers Christine Grammas of Technical Personnel Services, Inc.

Susan Lloyd of Source Services finds that "Most companies are willing to do almost anything for someone who is good. Before, candidates had to sell themselves to the companies. Now, they have to be sold—they're not doing the selling. On a daily basis, the professionals I deal with have offers that include up-front bonuses, telecommuting options, flex-time options, equity in the company, additional training, child care, tuition reimbursement, and additional vacation."

This is just a bit of the good news for high tech job seekers in Massachusetts, found in an article entitled *Job Seekers Can Name Their Own Ticket*, by Joanne M. Coletta-Levine of CNC News: " Talented professionals can negotiate a competitive salary, and also an agreement that includes such perks as up-front bonuses and work schedules that better suit their lifestyle. The current economy and low unemployment rate have created not only a huge surge in technical job opportunities, but also have made it a candidates job market."

If you have the requisite skills and you select a company that needs those skills, you are in a position to command a good salary and many life–enhancing extras as well. Make sure you do your homework. Find out what your prospective employer's company and similar companies in your area have been willing to do to persuade talented candidates to sign on the dotted line. Employee benefits today include much more than just medical insurance and 401(k) options. The high tech job seeker with solid skills should expect a very attractive package of benefits in exchange for coming on–board.

One of the myriad benefits of your application can be your geographic location. At the Tech Lease Website (<http://www.tech-lease.com>) the phrase "Hot Jobs" has a double meaning. With the exception of positions in Colorado and Massachusetts, the majority of the openings are in the sun belt; many are in Florida. That hardly means there are no jobs to be found elsewhere, but in high tech, the sun belt has been an area of exceptional growth. So unless you're a snowboarder (in which case, head for Colorado or Massachusetts), you could be able to find the job of your dreams, and live and work in a climate that was once reserved only for the comforts of retirement.

High tech is everywhere. You don't need to move if you're living in the house your great grandmother was born in, and the one you intend to die in, but there are certain areas of concentration within the high tech industry. The following excerpt is taken from an article about the high tech job market in the Boulder, Colorado area entitled *High-Tech Future Has Arrived For State's Best Paid Employees* (by Jerry W. Lewis, found at <http://www.bcbr.com/mar97/lewis2.html>).

"We don't have the largest number of high-tech workers. That's California, with 669,349, followed by Texas with at 313,460. And we don't have the highest average high-tech wage. Washington, home of Bill Gates and Microsoft, grabs that notoriety, at $57,555 with New Jersey second at $55,970. But Colorado receives high marks in a new study of "cyberstates" by the American Electronics Association. And according to Ted Lacina, the group's Mountain States Council director, gives Boulder County's sound manufacturing base much of the credit. According to The AEA, Colorado was one of ten states that added thousands of high tech jobs between 1990 and 1995. Others on the list may surprise you, for example, Nebraska, Georgia, and Utah. Colorado tied with Massachusetts for second place for the highest concentration of high technology employees—75 jobs per 1,000 private sector employees. Only New Hampshire was higher, with 78 per 1,000. Colorado came in third, behind Texas and Georgia for high tech employment growth in the five-year period. High tech firms created an astounding 22,199 new jobs, a 24% increase to a total of 114,005. Total payroll hit a whopping $5.4 billion, with an average wage of $47,067— 76 percent more than private sector workers."

The following passage from a job description posted on the Tech-Lease Web Site is typical of most current high tech job listings:

Project Manager, Benefits Software Development

Job Description: Fast growing company involved in benefit administration for other large companies is seeking a Project Manager for oversight and development/modification of software for benefits administration packages. Project planning, communications, documentation, reporting, modification, cost management and strategic planning are all a part of the duties involved. Develop prototype and coordinate with other ongoing projects. Would manage project processes from concept to development and implementation. Requirements: Will lead project's software development/software implementation. Desire some knowledge of health and insurance benefit packages and/or experience with huge payroll system. Perhaps some HR Human Resources experience. Candidate would oversee the global picture and must enjoy detail. Some Oracle experience helpful. Positive team player with persistence.

Most employers want to hire people who can follow a project through from its inception to its completion. In addition to the requisite computer skills (they are assumed in most instances), there are business skills that are expected from the ideal high tech employee. In the previous listing, the phrase "Perhaps some HR Human Resources experience" may seem mild. But a candidate who "would oversee the global picture and who must enjoy detail" as well as serve as a "positive team player with persistence" would clearly require extensive business, and perhaps managerial experience.

Still, out of two dozen "Hot Jobs" on the Tech-Lease site, only seven listed "people skills" among the job requirements, and none made mention of cultural sensitivity or diversity. Clearly, in high tech, the "tech" is the primary focus. As always in the technical field, the watchword is keeping up with new technologies. For instance, a Florida software firm, recently advertised for a technical position and closed with the following "requirement"—"to obtain those skill sets which he/she does not already possess."

Creative Pestering

Before we move on, let me leave you with another important piece of job search advice. Your taking responsibility for the job search makes all the difference in being successful. I call the most effective method of taking responsibility—rather than ceding it to the personnel department—"creative pestering."

Pester? Is that really the word we want to use? Well, yes, but there is a caveat. By "pester," I don't mean "be a pest," but rather "continually find ways to get yourself onto people's to-do lists in the nicest possible way."

Creative pestering *doesn't* mean—

- Leaving indignant or annoying messages on voicemail systems;
- Taking an attitude with receptionists and administrative support staff;
- Appearing combative or adversarial during the interview;
- Pulling crazy stunts that are likely to get you thrown out of the joint (like the overbearing applicant who decided to camp out in the president's office until the Top Banana agreed to meet with him personally);
- "Wearing down" key decision makers by declaring some kind of personal vendetta against the target company.

> **KEY POINT**
>
> The applicant who finds the most creative, persistent ways to pester is usually the one who gets the offer.

Creative pestering *does mean*—

- Doing the research necessary to make valuable proposals and suggestions—for free—that benefit the hiring official;

- Following every apparent "no" with a question about future hiring patterns, and staying in touch with decision makers who've turned you down to ask about new hiring initiatives;

- Proposing intelligent part-time contract assignments on your own initiative;

- Offering to buy decision makers an early breakfast to get the latest information on hiring within the company or industry;

- *Making countless pleasant phone calls* (that "pleasant" part is vitally important) in order to develop or learn more about professional opportunities.

The model resumes that appear in this book will serve to forward your candidacy in a dramatic way. When combined with a strategy of creative, persistent pestering, they will help you land that great job you deserve.

2
Some Resume Basics
What the Resume Is Meant to Do

The greatest truths are the simplest.
J.C. AND A.W. HARE

The reason most resumes don't fall into the WOW category is simple. Most people who write resumes have serious misconceptions about what a resume is really meant to do. What is the purpose of this ubiquitous document? What's it meant to accomplish? What are one's objectives in writing one? What *shouldn't* one expect a resume to do? Once you resolve these questions, you'll be in a much better position to craft a resume that makes the decision maker stop and say, "WOW!"

What a Resume Isn't

A resume is not an application for a position. Prepare yourself for a shock: The vast majority of resumes are completely ignored. In fact, so many resumes are sent "blind" to decision makers that the act of popping one in the mail, and doing nothing beforehand or afterwards to advance your career, amounts to little more than an exercise in wishful thinking. That brings us to (sigh) the bubble burst of the day: *Putting your resume into the mail does not, in and of itself, represent a meaningful form of outreach to a potential employer.* In fact, a fair number of experienced career counselors warn strongly that one's resume should never be mailed, *period.* Instead, they counsel that sending a superior letter (like the models that appear in Chapter Three) is a better initial form of outreach for making—and eventually telephoning—new contacts.

KEY POINT:

No matter what you may read or hear to the contrary from the prospective employer, you really shouldn't think of yourself as having "applied" for a job if you haven't talked to someone within the company about it. For most employment settings, the *don't-mail-the-resume* advice is pretty solid. You're far better off putting together a punchy written appeal that makes your "cold call" to a decision maker at a target company just a little warmer—or by calling active members of your own network and developing new contacts and leads in that way. To paraphrase the Beatles, mailing out resumes without any personal contact is a bit like trying to get a tan by standing in the English rain.

Once you've established some kind of relationship with the contact at your target company, *then* you can make arrangements for an in-person meeting, formal or informal. That's the time to pass along a version of your resume that is applicable to the opening in question. And yes, a well-structured letter—like the ones that appear in this book—can help you overcome the "send-in-your-resume-and-then-we'll-talk" trap so familiar to people who make job-related networking calls. All the same, I recognize that the temptation to "send a resume in and see what happens" can be incredibly strong. Because there are a few (and I do mean very few) situations where you may be forced to send a "blind" resume and cover letter, you'll want to take a look at the discussion of these situations that appears in the next chapter.

A resume is not a single document you can write once and consider "finished." Please don't make the (time-consuming) mistake of believing that a job search consists of developing a single resume, finding advertised openings, and mailing out copies of your resume until something happens. Instead, get on the phone with friends and associates, send out letters to new prospective employers (see Chapter Three), and target each resume you send out. By doing so, you'll set yourself apart from the pack. It's certainly true that not many people actually like writing resumes. Most people want to write the resume once and mail it to fifty different employers. That probably explains why these documents so rarely have any direct bearing on the employment openings they're supposed to help applicants track down.

The resume you pass along to a decision maker should be focused specifically on the company and/or position in question. A fair number of the sample resumes that follow in this book take the perfectly reasonable, and entirely accurate, step of listing as an "Objective" the specific job at that company that the job seeker is pursuing. This is a much sounder course of action than stating your objective from your own perspective (i.e., "Locate a position within a dynamic firm that will allow me to grow professionally.") That you should be ready to use your computer to customize all essential elements of your resume's text is taken as a given here. Learn what problems the decision maker is hoping to resolve, and then focus your resume on those problems!

A resume is not an affidavit. It's an advertisement, a marketing tool, a device that must use every (accurate!) statement it possibly can to forward your candidacy.

Many job applicants, and particularly those seeking high tech-related positions, make the mistake of slavishly following preestablished formats or patterns to pass along the "right" information. The "right" information is that which inspires the decision maker to pursue your candidacy further—and any sample resume you encounter, including those that appear in this book, should be considered a suggested format for adaptation to the specific situation you face.

What A Resume Is

A resume is something meant to be scanned. Long, dense blocks of type have a way of turning hiring officials off although there are certain situations where they can serve you well by amplifying points specifically raised by the prospective employer earlier in the process.

Most (though not all) of the resumes that appear in this book feature condensed "talking points" rather than long essays on particular aspects of one's work experience. There's a reason for taking this approach *virtually no one reads resumes,* at least not in the early stages of your contact with the target company. Expect your resume to be scanned quickly, not studied minutely.

A resume is a convenience for the hiring official. Shocking news! The resume is really a way to screen you *out* of the organization. That's right—the decision maker typically uses your resume to make more or less instantaneous "yes-or-no" determinations about your candidacy—and the candidacies

KEY POINT

It means you may have only a few seconds—perhaps a few *fractions* of a second—to make or reinforce a positive impression with the resume you create.

of literally hundreds of other applicants. There's almost always a big pile of resumes from potential applicants. The hiring official has to make that pile disappear, and thus almost always uses resumes to find out what doesn't match his requirements, thereby saving time that would otherwise have to be devoted to countless face-to-face discussions with eager applicants.

You have to make sure your "grabber" introductory material really does grab! And, since people tend to zip to the bottom of the page, you have to make sure that the "Hey, look me over" text you select to close out your resume reinforces the message, "This one looks interesting." Within the body of the resume, you have to isolate benefits, and make the reader awfully curious about finding out more in a hurry! And speaking of curiosity ...

A resume is an opportunity for you to leave the reader wanting to learn more. Insofar as the resume serves as a "silent spokesperson" in your absence it must, like any good advertisement, eventually inspire your audience to action. In most cases the outcome you're after is a simple one: You want the person to pick up the telephone and either offer you the job or ask you to come in and discuss the job in detail. The objective is not to supply a full-blown professional biography, but to supply *enough telling facts* to make a decision to contact you easy to justify. After all, you'll want to save some of the heroics for your face-to-face discussion with your contact!

Now that you've gotten a good idea of what your resume is supposed to do, you're ready for ...

Twenty Big Questions That Will Help You Develop Your "Wow" Resume

Here are twenty questions to ask yourself about your own work history and education. Find a place where you can devote *at least ninety uninterrupted minutes* to the development of written answers to each of these questions. That's five to seven minutes per question.

Jot your answers down in a looseleaf notebook. Give your "best guess" answers for now. Be honest, and don't let yourself get bogged down in technical detail. If you find yourself thinking that answering a question fully requires some in-depth research on your part, leave that aspect of the question blank for now and make a note to come back to it later. For now, get the broad outlines to your answers to each of the Big Twenty.

1. What are the three most dramatic examples of verbal or written praise you've ever received from any supervisor or client? What awards, commendations, or formal acknowledgments have you received in any work setting?

2. What was the most recent job you held? (If you're presently employed, this will be your current job.) List the most important duties?

3. What skills do/did you have to develop to deliver superior results on the job in that environment?

4. Think of at least three situations when *failing* to do something you do/did in that work environment would have resulted in disaster for your employer. How much money/time/resource would have been required to rectify the problem?

5. Think of at least three times a supervisor in this position outlined a problem for you to solve—a problem you *did* solve successfully. What was the positive outcome of each of those solutions? Did you make a system or procedure run more smoothly? If so, how much more smoothly? Did you save money for your employer? If so, how much money? Did you save time? If so, how much time?

6. Think of at least three instances when you personally "saved the day" as part of your work on this job—perhaps by thinking quickly in an emergency or acting responsibly during trying circumstances. Did a computer ever crash—leaving you to pick up the pieces? Was a customer ever angry—over something you were able to resolve? Was a deadline ever moved up to a date that seemed impossible—but wasn't?

7. Answer question 2 for your next most recent job.

8. Answer question 3 for this job.

9. Answer question 4 for this job.

10. Answer question 5 for this job.

11. Answer question 6 for this job.

12. Answer question 2 for the job you held before that.

13. Answer question 3 for this job.

14. Answer question 4 for this job.

15. Answer question 5 for this job.

16. Answer question 6 for this job.

17. Write down the specifics of at least five situations where coworkers came to you for help and you were able to provide it. What was the worst-case scenario that was averted by your taking action? What positive outcome emerged instead?

18. Make a list of any extracurricular activities you pursued in school that provided you with experience that either directly related to the high tech position you'd like to win—or required you to develop significant leadership skills.

19. Make a list of any charitable activities that directly relate to the high tech position you'd like to win.

20. Make a list of at least fifteen people—former employers, colleagues, professional associates—who would be willing to develop short written recommendations for you or agree to endorse a recommendation you composed.

As you peruse the resumes in this book, you'll find that most of them make use of the kind of information the questionnaire above asks you to develop. By taking time now and devoting a good, solid 90 minutes to answering the questions above in writing, you'll put yourself in a position to highlight what employers really want to see: potential answers to pressing problems!

In the resume you develop using the models that appear in this book, you'll be able to provide the potential answers to those pressing problems in the form of ...

- Compelling (legitimate!) endorsements from third parties
- Examples of performance that saved time or money or increased the efficiency of your company
- Instances when you took the initiative and forestalled disaster

Those are the kinds of resume elements that make for WOW resumes. There are dozens of examples of these types of items in the pages that follow—but take the time now to develop the material that puts you in the best possible light.

There Are References and There Are References

By the way, I should note here that some resume-writing authorities take a dim view of attaching written references to your resume on separate sheets of paper. I supposed I'd have to agree—this tactic looks a little desperate, and it's all too easy for the separate endorsements to become separated from your resume. (My experience is that a good many managers resent anything that adds to the "clutter factor" on their crowded desks.) That's not the same thing, however, as *incorporating* the highlights of written or spoken endorsements, which can be a very powerful tool indeed.

In Chapter Five, you'll find a couple of brief, helpful checklists—summaries of resume and employment letter commandments, if you will. Those lists will help you evaluate your written appeals and make them as sharp as they can possibly be. For now, take the time to answer the Big Twenty Questions in detail. That work is the foundation of any attempt you'll make later to develop a superior resume based on the samples in this book.

Please do not continue on to the next portion of this book until you've completed the questionnaire work in this chapter!

Use Resume Models for Job Search Success

In this book, there are dozens of examples of superior resumes that will help you take the raw material you can develop by means of the above questionnaire to develop your own WOW resumes. As you examine them, be prepared to adapt elements from more than one resume. Use what works. Use what makes sense for your situation.

KEY POINT

Adapt the formats of appropriate resumes—whether they follow a chronological format that examines dates of employment, a functional outline that emphasizes key skills of interest to *a particular reader*, or a combination of the two—with an eye toward making a dramatic, confident, positive impression on your reader.

When You Get Stuck

It happens to the best of us. Sometimes you run dry while you're working on your masterpiece of aggressive employment self-promotion. Don't panic. Find a constructive way to get around your roadblock.

Here are two strategies to consider when you find yourself staring at that long, blank sheet of white paper while setting up your initial notes—or, if you've finished your initial questionnaire and are transferring your work into a more polished form on your personal computer, while that insistent cursor keeps blinking whether or not you've got something interesting to say.

Strategy One

Take a short break. Get up for a minute or two to grab a drink of water—or pop a high-energy tape or CD into the nearest music system. Most cases of "writer's block" come when we're pushing ourselves unreasonably, trying to turn out something fresh when our minds have been on the case too long. As long as you don't use the "drink-of-water" routine as an excuse to keep your stretches of constructive work from disintegrating into a four-minutes-on, two-minutes-off charade, you can usually get further by allowing yourself a modest break, for instance, between fifteen- to twenty-minute time slots spent working on your resume. When you hit a wall, find a way to clear your mind. Don't just keep demanding new results from the same exhausted brain cells. (Also effective, and worthy of consideration if you've just taken a short break, move on to another section of your resume, and a little later on come back to the problem that has you stumped.

Strategy Two

Use a Superstar Verb for inspiration. The most powerful parts of your resume will probably be sentences that begin with active, results-oriented verbs. ("Maintained a 99.85% error-free contact database for use by sales force.") Here's a list of over 130—play a little game with yourself and find ten sentences that describe your work background, beginning each with a verb that addresses the area where you're stuck.

Superstar Verb List

accepted (responsibility, heavy workload, challenge, etc.)	accomplished
acted as troubleshooter for ...	acted to
adapted	adjusted
administered	advised
allocated	analyzed
appraised	approved
arranged	assembled
assigned	audited
authored	authorized
balanced	briefed
budgeted	built
calculated	catalogued
chose	clarified
coached	compiled
computed	conducted
consolidated	convinced
coordinated	critiqued
customized	cut
decreased	demonstrated
designed	determined
developed	devised
diagnosed	directed
dispatched	drafted
edited	eliminated
established	estimated
evaluated	executed
explained	facilitated
forecasted	formulated
found	founded
headed up	hired
identified	implemented
improved	increased
informed	initiated
inspected	installed

instituted
interpreted
invented
learned
led
managed
monitored
negotiated
organized
overhauled
planned
prepared
produced
promoted
publicized
qualified
reached out
reconciled
recruited
researched
reviewed
rewrote
scheduled
selected
shaped
simulated
solidified
specified
strategized
strengthened
supervised
systematized
trained
turned around
wrote

instructed
interviewed
launched
lectured
limited
marketed
motivated
operated
originated
oversaw
predicted
prioritized
programmed
prospected
published
reached
recommended
recorded
redesigned
resolved
revitalized
saved
screened
set
showed
sold
solved
spoke
streamlined
summarized
surveyed
toured
trimmed
upgraded

3

Supporting the Message

The Lowdown on Letters

Invention breeds invention.
RALPH WALDO EMERSON

Many people are surprised to learn that the "look and feel" of one's written job search material can have a surprisingly powerful impact on the employer's final decision. Rest assured—it's true!

I strongly suggest that you coordinate your written job search correspondence by choosing a particularly, striking set of stationery and using it for all your correspondence with target companies. The paper you choose doesn't have to be expensive or visually overwhelming (in fact, it probably shouldn't be either), but it should broadcast *professionalism* in a consistent, understated way.

When the time comes to make a final decision about a new hire, the odds are good that your contact within the organization will make a final review of the written correspondence received from all applicants. How much do you think it will aid your cause for that decision maker to see that your resume, and the series of letters that accompanies it, all share the same tasteful color scheme and basic look? This may seem like a small consideration; but believe me, paper selection can make a difference. (So can the decision to type or word process all your letters if your handwriting isn't exactly the world's neatest.)

I've already spoken a little bit about the dangers of mailing your resume. Let's examine this point in a little greater depth now as it relates to the alternative we'll be discussing in this chapter—pre-resume letters.

There are three main points to bear in mind when it comes to popping your resume into the U.S. mail. They are:

1. Secretaries and administrative assistants routinely "screen out" unsolicited resumes. As a general rule, they're paid to do so! That means that *no matter what you say in the cover letter,* there's a very good chance that someone other than your intended recipient is going to spot the resume and put your correspondence aside for "later review." You know how often you get to the business correspondence you set aside for "later review." That's about how often hiring officials will make it to scan your resume.

2. Secretaries and administrative assistants virtually never know the difference between a resume in which someone has asked for an appointment and a resume that's been sent "cold." They may spot your "as requested" notation—then again, they may not.

3. Secretaries and administrative assistants don't screen most personalized business-related correspondence, i.e., if your correspondence does not include a resume, it stands a much better chance of making it through to its intended recipient, especially when you're contacting employers about something other than advertised openings.

What does this all add up to? *Statistically speaking, mailing resumes "blind" is more likely than not to be a waste of time, effort, and money.* On the other hand, sending a broadcast letter—a souped-up letter that offers more pertinent detail about your career than an untargeted letter might, but less than a resume would—might ensure that a decision maker actually sees what you've written.

Are there *some* situations in which you'll be well advised to send a resume and a top-notch cover letter? Sure. The hiring official may be located in another city, and may demand to see a resume before asking you to fly out to discuss an opening. Or the hiring official may have lost the original copy of the resume you delivered. Or the hiring official may have asked you to develop a new version of your resume—and may simply refuse to meet with you until he or she has had the opportunity to review it.

All these possibilities exist, and in these cases you'll need to assemble a superior cover letter, one similar in impact to the samples that follow. But in *most* situations, your best course of action will be to make virtually any excuse

KEY POINT

If you do opt to send a resume through the mails, *never* send one without a sharp-looking cover letter on matching stationery.

that allows you to *speak* to your contact before personally delivering your resume, which should be targeted *directly* to the needs, requirements, and specifications of the employer with whom you've spoken.

One more important word of advice—think twice before you get overly "creative" when it comes to establishing the means of connection between your contact person and your resume or letter. No doubt this will be familiar advice by now, but it bears repeating: Try your level best to find some way to hand the package over yourself during a face-to-face meeting. Such a strategy will be far more effective than coming up with some elaborate delivery system designed to get the decision maker to "pay close attention" to your resume.

Overnight delivery services—bicycle couriers—flowers—gifts—oversized boxes—registered mail. ... Believe me, experienced hiring officials have seen it all, and they're more likely to wonder what you're trying to compensate for rather than being impressed by your ingenuity. Take the money you were going to spend on renting a marching band or a small jet plane, and put it back into your bank account. Set aside another hour or two to make sure the material that's on your resume tells a dramatic, confident story—in a way that's immediately realized by the reader.

The "Perfect" Letter

So—how do you use written materials to pave the way for a relationship with a contact within your target industry? Here's the single best tool I know of. It may not be a "perfect" letter, but it's pretty darned close. It shows that you've done your research, and that you're willing to take on responsibility for cheerleading your own candidacy through the organization. It should be no surprise to you by now that I advocate sending this letter without a resume and following up with a phone call; however, you should know that the model below can also serve as a superior cover letter on those rare occasions when you need to use it as one. Just add the sentence "My resume is enclosed." to the beginning of the paragraph before "Yours truly."

(Date)

John Miller
ABC Medical Equipment Center
456 Main Street
Mytown, State 00000

Dear Mr. Miller:

Jane Owens, in your Human Resources Department, tells me that you're looking for a Senior Programmer.

ABC's Requirements	*Jane Smith's Experience*
College degree	Bachelor's degree in Computer Science from Worcester University, 1984
Two years experience with database management	Three years experience with database management at Priparm Computer Associates, and six years at American Widget
Four years experience in an industrial environment	Six years experience in a service environment with American Widget
Knowledge of all industry-specific programming languages	Knowledge of ANSI COBOL, CICS, VSAM, OS/DOS, JCL, ASSEMBLER, PASCAL AND IMS
Familiarity with appropriate software applications	Nine years experience with spreadsheet programs as Excel and Lotus 1-2-3

I think we should talk about the ways I could make a significant contribution to ABC Corporation. I look forward to speaking with you soon.

Yours truly,

Jane Smith
123 Main Street
Mytown, State 00000
(555) 555-5555

P.S. I will plan on calling you at 9:00 A.M. on Tuesday, July 16th. If this is not a convenient time for you, please have Robin give me a better time to return the call.

This simple letter—which requires, as you've no doubt noticed, that you do a little phone research to determine the nature of the open position and the name of the receptionist, secretary, or administrative assistant—is the single most effective weapon in your arsenal when it comes to making contact with decision makers. It does not guarantee to get you an enthusiastic return call—no letter will always do that. But it is virtually certain to capture the interest and attention of anyone who reviews the mail perhaps once a week, and has a vague appreciation of the fact that his or her organization is looking to hire a computer programmer.

When you follow up by phone as promised, you're likely to hear a pause and the rustle of paper as your contact searches for that intriguing letter that crossed the desk a few days back. That's what you want to hear. When your contact fishes the document out of the pile, find an opportunity to ask repeatedly and politely for an in-person meeting to discuss the position—and, if time allows, ask the person what he or she "is trying to get accomplished" in the area you're interested in. Odds are you'll get some meaningful feedback—and some inspiration about the best way to target your resume for the meeting.

The beauty of the two-column approach is that you *focus only on matches between what you offer and what the prospective employer wants.* You don't bore the reader with your life story, and you don't supply lots of details that don't fit into the "right profile" the hiring official is responsible for tracking down. But suppose you don't have any idea about the requirements of the position? Suppose you're doing all the right things—avoiding fixation on the classifieds, isolating fast-growing employers, devoting a certain period of the day to cold calls to managers—and you *don't know* whether there's an opening that's right for you at a promising company? As luck would have it, you can still use a letter to your advantage. Consider the two samples that follow—each of which can, like the one you just saw, be adapted to those infrequent situations when you must pop a resume into the mail.

(Date)

Mr. John Miller
ABC Corporation
456 Main Street
Mytown, State 00000

Dear Mr. Miller:

According to *Business Week,* the American widget industry is poised for "significant expansion into Eastern European and South American markets" within the next two years (*Business Week,* January 16, 1998, page 123). I applaud the groundbreaking work your firm is doing to expand into these new markets, and I would love to help contribute to your firm's steady growth.

I gather from the *Business Week* article that your company is among those attempting to significantly increase business in those areas of the world. If that's the case, you're likely to need help managing your computer networking systems. I offer you:

> Nine years of superior high-efficiency performance as a LAN Manager in the widget industry
>
> Superior knowledge of WidgetManager and other industry-specific programs
>
> A team-first attitude and the ability to work both independently and as part of a work group

Since I will be in your area next week, I would like to visit with you so we can discuss how we might work together to increase efficiency and reduce overall costs through top-notch network management at your company.

Sincerely,

Jane Smith
123 Main Street
Mytown, State 00000
(555) 555-5555

P.S. I will plan on calling you at 9:00 A.M. on Tuesday, July 16th. If this is not a convenient time for you, please have Robin give me a better time to return the call.

By the way, my experience has shown that the P.S. is the very best place to serve notice as to what you plan to do next. The more you're able to customize the P.S.—for instance, by making reference to a secretary or assistant—the better off you'll be. And if you *reach* that secretary or assistant, don't get haughty or start ordering him or her around. Treat this person *exactly* as though he or she were the Big Boss.

Here's another example of an eye-catching cover letter that paves the way for a personal phone call. This one does the trick by highlighting a single example of superior performance that leaves the reader wanting to know more.

(Date)

Mr. John Miller
ABC Corporation
456 Main Street
Mytown, State 00000

Dear Mr. Miller:

I was named "Data Analyst of the Year" at my company's national awards dinner last year because I made major contributions to the creation of an entirely new, automated order entry system after only six months on the job.

I'm a computer-savvy, goal-oriented team player who follows the widget industry closely. The January 16, 1998 issue of *Business Week* leads me to believe that your company plans to expand in the future—and that you'll soon be needing a qualified data analyst like me.

Let's meet to discuss the possibilities!

Sincerely,

Jane Smith
123 Main Street
Mytown, State 00000
(555) 555-5555

P.S. I will plan on calling you at 9:00 A.M. on Tuesday, July 16th. If this is not a convenient time for you, please have Robin give me a better time to return the call.

As you've probably already gathered, I'm a big believer in the "less is more" principle when it comes to job search letters. (I think the same basic principle applies to resumes, by the way; I've met too many hiring officials who hate reading multipage resumes to feel otherwise.) The aim, after all, is not to drown the (probably overloaded) reader in new facts, but to elicit that simple "WOW" reaction that leads naturally to the question—"When can you come by for an interview?" I've talked to plenty of hiring officials who felt overwhelmed by a resume that left little to the imagination ... and, often, little to discuss with the applicant.

Here's one more effective employment letter that you may wish to use as a model, one that's based on a previous phone conversation between you and your contact. Note how it takes the bull by the horns—that is, takes responsibility for a dramatic "next step"—and how it employs effective appeals based on points raised during your call.

(Date)

Mr. John Miller
ABC Corporation
456 Main Street
Mytown, USA 00000

Dear Mr. Miller:

It was great to talk to you recently concerning employment opportunities at ABC.

I was particularly intrigued by what you said about the new widget outsourcing program, which I understand needs to be up-and-running very soon. With your permission, I'd like to put together an outline of an agreement that would allow me to help ABC develop a detailed plan of attack for your project, on an "independent contractor" basis—including recommending vendors and developing training materials for the new hardware and software you'll need.

I plan to fax this outline to your attention this coming Friday, and I'll follow up by phone on the following Monday.

Thanks for taking time to speak with me.

Sincerely,

Jane Smith
123 Main Street
Mytown, State 00000
(555) 555-5555

P.S. I will plan on calling you at 9:00 A.M. on Tuesday, July 16th. If this is not a convenient time for you, please have Robin give me a better time to return the call.

Talk about creative pestering! The letter above provides an excellent example of exactly what that looks like when it's committed to paper. Did your contact volunteer anything about your working as an independent contractor for ABC? No—but there's certainly no reason for you not to bring the subject up, especially if you have some insight about where the company is going in the near future. Did your contact say anything about talking to you about your outline on Monday morning? No, but simply faxing the material would probably lead to a long silence. What's the harm in following up persistently and politely?

Now it's time to take a look at the stars of our show— the model documents that will help you craft your own WOW resume. As you make your way through the main section of the book, remember that you're going to be talking to more than one potential employer over the course of your job search. That means you'll need more than one resume. Don't wed yourself to a single model from the examples that follow. Pick the format that provides the best "fit" with your situation, and customize it to the needs and interests of the person and organization you're targeting.

4

Don't Send It Off Yet!

Double Check Everything

Chi Wen Tze always thought three times before acting. Twice would have been enough.
CONFUCIUS

No matter how excited you are about the opportunity you're pursuing, don't consider your resume and cover letter complete until you've consulted the following checklists—presented to you as "Commandments."

There are plenty of resume horror stories making the rounds. I've heard tell of resume writers who committed grievous spelling errors in the very lines in which they boasted about their attention to detail; resume writers who focused on catastrophes at work that they vowed not to repeat, and resume writers who let ludicrously inappropriate word choices torpedo their chances for getting a good job. Don't let that happen to you! Follow these Commandments for Perfect Letters and Commandments for Perfect Resumes.

TEN COMMANDMENTS FOR PERFECT COVER LETTERS

Customize Your Document to the Intended Audience.

Focus Your Remarks within a Few Concise Paragraphs, and Don't Exceed a Single Page.

Read Your Letter Carefully and, if Possible, Subject It to a Computerized Spell Check.

Then, in Addition to Your Computerized Spell Check, Enlist a Trusted, Literate Friend to Review the Text for Spelling or Style Errors.

Never Bring Up Salary Unless Instructed by the Contact to Do So, in Which Case You Should Discuss Broadly Scaled Salary Ranges.

Never Focus on a Negative Element of Your Background.

Use the Word "I" with Restraint.

Include Full Contact Information in Your Letter.

Close Your Letter with a Promise of, or a Request for, Future Action.

Always Tell the Truth.

TEN COMMANDMENTS FOR PERFECT RESUMES

Customize Your Document to the Intended Audience.

Don't Bore the Recipient.

Read Your Resume Carefully and, if Possible, Subject It to a Computerized Spell Check.

Then, in Addition to the Computerized Spell Check, Enlist a Trusted, Literate Friend to Review the Text for Spelling or Style Errors.

Include Only Facts That Support Your Cause—Never Confuse Your Resume with a Confessional Document.

Display Energy, Creativity, and Personality without Exceeding the Bounds of Good Taste and Professionalism.

Break Your Points Up into Readable Chunks.

Eliminate Unnecessary Fluff and Trivia.

Include Full Contact Information in Your Resume.

Always Tell the Truth.

Review these lists, check them twice, bear them in mind as you compose your written appeals—and your campaign to land the high tech job you deserve will have been well and truly launched.

Good luck!

5

The Resumes

How to Start Your Resume With a Bang

To have begun well is to have done half the task.

HORACE

There are plenty of models presented for your own high tech resume, featuring a wide variety of educational backgrounds, levels of professional experience, and text styles. Scan them by title, but keep an eye open for style or structure decisions worth adopting for your own resume, even if the job title is radically different from the one you are pursuing.

A strong personal statement and use of quotes—references lead off this resume, bringing attention to past accomplishments and highlighting personal qualities.

Mary Smith
45 Evansdale Drive
Anytown, STATE
(555)555-5555
E-mail: smith@network.com

WHAT I OFFER THE RIGHT ENGINEERING/COMPUTER SERVICES FIRM
Proven analysis and project management strengths, computer hardware/software knowledge, and teaching and customer service abilities.

WHAT MY REFERENCES SAY
"A take-charge, results-oriented attitude that has resulted in top-notch project management and critical revenue contributions to the organization." Fred Beatty, Senior Vice President, Amtech Corporation.

"Your solutions resulted in a 24% increase in productivity in our department. Many thanks!" Ellen Woodland, Proprietor, Reporting Systems Hardware.

"A one-woman technological solution machine who has risen to the occasion in any and every setting where we have worked together." Vera DiGregorio, Partner, Phoenix Solutions.

LEARNING EXPERIENCES
1990–1994 B.S. Computer Science, Redlands University, San Rafael, California; awarded highest honors Dean's list, four quarters

ON THE JOB
1997–1998   **Partner, Phoenix Solutions**

Co-founded this consulting partnership dedicated to resolving hardware and software problems for area manufacturing companies. Personally handled on-site analysis, inspection, and implementation of proposed solutions for fourteen corporate clients.

Conducted in-person training in a wide variety of settings.

Assumed 24-hour on-call responsibility for key projects during implementation phase.

1994–1997   **Senior Technical Analyst, Amtech Corporation**

Used skills in hardware and software design and troubleshooting to assume lead role in developing six new products for this fast-growing corporate data management software firm. Headed up design efforts for successful MaxInfo 1.0 package, responsible for $2.4 million in revenue during fiscal year 1994. (Amtech was acquired by General Technologies in 1996, leading to restructuring campaign.)

Chosen to develop training programs for the firm's eleven most important customers

Researched, prepared and developed code for companywide reassessment of network systems, leading to "far more streamlined product planning effort," according to senior management.

YES, I SPEAK . . .(COMPUTER LANGUAGE INFORMATION)
Programming: JAVA, C and C++, ASSEMBLY, COBOL, PASCAL, BASIC, ORACLE, UNIX
Software: MS/Word 6.0, Lotus 1-2-3, Windows 95, Act!, Corel Draw, and many other programs.

Extensive military experience, outlined concisely and powerfully in the first
paragraph of the Professional Summary, is the focus of this resume.

John Smith
45 Evansdale Drive
Anytown, STATE
(555)555-5555
E-mail: smith@network.com

OBJECTIVE

A position where my skills acquired in the military and through formal education would add measurable
value to Acme Systems Corporation.

PROFESSIONAL SUMMARY

UNITED STATES NAVY August 1993–December 1995: Personnel Support Detachment (Nov.
1994–Dec. 1995) Activities centered on the processing of a large number of military personnel dis-
charges. Operated computers and updated medical records, inputting data and closing out files as part
of routine business.

USS Carl Jones Aircraft Carrier (CVN 70) Shipboard Airman (Aug. 1993–Dec. 1995). Homeport:
Alameda Naval Station, California. Assisted with On-deck Flight Operations of F14, FA18, A6, EA6B,
E2 Hawkeye, S3 Viking. Originally assigned as Crash and Salvage firefighter and troubleshooter and
Plane Handler. Participated in simulation drills as part of a two-man team, checking for hot spots.

Completed Recruit Training Command at Great Lakes, Illinois training base.

Military Training Summary - Shipboard Aircraft Fire Fighter- CBR Warfare Defense - Advanced Fire Fight-
ing & Rescue Operations- Basic Lifesaving Qualifications - Damage Control- Expert Vehicle Operator.

CAREER RELATED EDUCATION

STAN SMITH COLLEGE, Greenvale, New York, B.A. in Communication and Audio Engineering,
1985. Extracurricular Activities: Radio Station WCWP: Ran programs and worked audio control board
as part of overall station operations.

ADDITIONAL EMPLOYMENT

Junior Manager, Accolit-Gil Ecclesiastical Arts, Southhold, New York (1989–1993). Supervised 3–8 general laborers on the job site and set up for work, renovating old Historical Churches. Mechanic/Inspec-
tor, Budget Rent-A-Car (1989) Safety & Emission Inspector, New York City Taxi & Limousine Commission (1989); Apprentice Mechanic/Inspector, Action Diagnostic Center, Whitestone, New York
(1988–1989).

References Available Upon Request.

A customized appeal to the target company in the opening lines of this
resume helps it to stand out from the competition.

John Smith
45 Evansdale Drive
Anytown, STATE
(555)555-5555
E-mail: smith@network.com

OBJECTIVE: To obtain a position as an Avionics Technician with ABC Engineering, either working on the line or the bench, where my education and training in Avionics Technology would lead to increased profitability and efficiency.

EDUCATION AND SPECIAL TRAINING

COLLEGE OF SCIENCE AND TECHNOLOGY, Dallas International Airport, Dallas, Texas: Associates in Occupational Studies, A.O.S./Avionics Technology, September 1992

Core curriculum included the following coursework:

January 1992 Semester, FCC LICENSE REVIEW - Prepared for and took practice tests. Application for next test date is on file.

ATC TRANSPONDERS (B) - Ability to troubleshoot and repair, remove and install systems.

DISTANCE MEASURING EQUIPMENT (B) - Ability to troubleshoot DME and other systems connected.

DEGREE PROJECT - MAINTENANCE: This involved the construction of a VHF Communications Filter, adhering to filter specific frequencies.

September 1991 Semester, VHF COMMUNICATIONS TRANSMISSIONS (A) - Studied air-to-air and air-to-ground transmissions and receiving systems. Worked with blueprints for whole circuitry layouts for various types of communications systems.

HF COMMUNICATIONS TRANSMISSIONS (B) - Studied radio waves. VHF NAVIGATION (A) - Studied VOR aircraft navigation systems. MICROPROCESSORS I & II (B) - Learned to repair IBM mainframes. Basic knowledge of programming applications, and the ability to understand the inside of the computer.

February 1991 Semester, POWER DISTRIBUTION and FLIGHT SYSTEMS (B) - Worked with entire aircraft electrical system layouts (737s and 727s). Utilized electrical charts and block diagrams.

ELECTRICAL CIRCUITS I & II (A) - Studied transistors, semiconductors, and circuits. Worked with diagrams as well as actual components.

September 1990 Semester, DIGITAL ELECTRONICS (A)

February 1990 Semester, INTRODUCTION TO COMPUTERS (A) - Studied and worked with software packages: Lotus, Database III, and WordPerfect in a business oriented environment for purposes of generating reports. Simulating business operations, Profits/Losses.

TECHNICAL MATHEMATICS I & II (A)

September 1989 Semester, TECHNICAL DRAWING (B)

CHEMICAL ENGINEER

This is an example of a brief resume that can be accompanied by a powerful, customized, one-page summary of relevant computer-driven research projects.

JOHN SMITH
45 Evansdale Drive
Anytown, STATE
(555)555-5555
E-mail: smith@network.com

PROFESSIONAL OVERVIEW
An efficient research and development professional skilled in computer modeling, process simulation, and chemical engineering laboratory work.

EDUCATION
Ph.D. 1995, The University of Regina, Saskatchewan, Canada, Department of Chemical and Petroleum Engineering
Specialized in Chemical Engineering.

EXPERIENCE
1995–1996 Postdoctoral Research Fellow and Sessional Instructor University of Regina, Saskatchewan, Canada, Department of Chemical and Petroleum Engineering
Taught undergraduate Statistics and Probability for Engineers.
1990–1995 Research Assistant University of Regina, Saskatchewan, Canada, Department of Chemical and Petroleum Engineering

LABORATORY EXPERIENCE
Deep experience with laboratory instruments.
Communicated effectively with mechanical and chemical engineering technicians and directed them toward key tasks.

COMPUTER EXPERIENCE
Skilled in a wide variety of platforms and applications.

TEACHING EXPERIENCE
Taught Mathematics, Statistics, Chemistry, Physics and Heat, and Mass Transfer.

PUBLICATIONS
Three

CITIZENSHIP
U.S. permanent resident

PROFESSIONAL AFFILIATIONS
Member of American Institute of Chemical Engineers

The applicant's choice to lead the resume with a review of specific software skills is a particularly good one here, given the traditionally heavy emphasis on "hitting the ground running" in the design environment. The ability to pick up a half-completed project may lead to freelance assignments that stand a good chance of turning into full-time offers.

JANE SMITH
Graphic Designer
45 Evansdale Drive
Anytown, STATE
(555)555-5555
E-mail: smith@network.com smith@network.com
To view my on-line portfolio, surf to www.janesmith.com

COMPUTER SKILLS

Adobe PhotoShop
Adobe Illustrator
CADD—AutoCad 12 through 14, and AutoLISP
Quark Xpress
GifBuilder
Micrografx Simply 3D
Microsoft Office (all applications, including PowerPoint, Microsoft Excel and Microsoft Word)
Web design with HTML and Javascript
Four years of Macintosh experience; three years of IBM experience.

DESIGN EXPERIENCE

BERSMAN, INC.   San Francisco, California (October '97–current)
Graphic Design Specialist

Developed packaging designs, promotional brochures, and original computer-generated images for Fortune 100 consumer products firms. Accounts included Deria Electronics, BeanWay Transit, Vereston Associates, and North Shore Snacks. Handled design, layout, and modification of company web site. (www.bersman.com); honored with "Best of Region" award from West Coast Web Developers Association, 1998.

Also experienced in trade show booth design and setup, development of direct mail marketing pieces, and catalog design.

CRESTWAY FOODS, INC.   Dallas, Texas (June '95–October '97)
Graphic Designer

Used advanced software tools to develop logos, food labels, and prototype package designs. Developed slide presentations on tight deadlines for presentations to retail marketing accounts representing $22 million in annual revenue to the firm.

SENCO TOURS, INC.   Houston, Texas (January '94–June '95)
Graphic Designer

Developed four-color layouts for this leading Texas adventure tour organization; set up flyers, postcards, catalogs, and brochures. Reported to President of company; delivered superior materials under very tight schedules.

EDUCATION

Bachelor of Fine Arts, May 1993
Houston College of Design, Houston, Texas
Major: Computer Design

COMPUTER REPAIR TECHNICIAN

Recent educational efforts are placed at the forefront of this resume and supported by carefully selected highlights of earlier work experience.

JOHN SMITH
45 Evansdale Drive
Anytown, STATE
(555) 555-5555 / E-mail: smith@network.com

EDUCATION

CHARLES RIVER TECHNICAL SCHOOL   Cambridge, Massachusetts
Received Certificate 9/94. Digital Computer Technology course curriculum included: Mathematics, Basic Electricity, 5/96. AC & DC Circuits, Power Supplies. Semiconductor Theory and Troubleshooting. Operational Amplifiers, Computer Mathematics, Boolean Algebra, Combination Logic Circuits, Flip Flops, Counter Registers, Memory Systems, D/A Conversion, Programming Concepts & Computer Troubleshooting, Introduction to Microprocessors.

PROFESSIONAL EXPERIENCE

TECHNICAL SCANNING INC.   New Haven, Connecticut
2/93–8/94 Field Service Trained on high-speed check processing unit, which included optical character reader, magnetic ink character reader, and ink jet printer. Also worked on peripherals such as 80-megabyte Pertec disk drives, line printers, and Harris PC6 controllers for sorters.

4/94 On-call field service in Boston, Massachusetts. Interacted with customers and maintained equipment (such as personal computers).

8/94 also serviced GA processors and bank teller equipment.

PARAGON CABLE   Lynn, Massachusetts
8/92–2/93 Bench Technician Head and base repair which included troubleshooting on modulators, voltage regulators. and transformers.

REFERENCES AVAILABLE UPON REQUEST

Here, an applicant interested in moving up to a field position uses
relevant retail experience very effectively.

MARY SMITH
45 Evansdale Drive
Anytown, STATE
(555)555-5555
E-mail: smith@network.com

OBJECTIVE

A position in an active sales environment where my ability to exceed sales targets and develop profitable new business relationships will be a significant asset to ABC Company.

EDUCATION

9/84–Present   ST. MARY'S UNIVERSITY, Richmond, VA
B.S. Degree - 1/90

Major: Computer Science
Minor: Business

Computer Courses: Basic Assembler, Pascal, Cobol, RPG PL1, Data Base III Plus, Slam II Lisp. Used IBM PC and Mainframe (Honeywell Multics System)

Business Courses: Accounting I and II, Marketing Economics I and II, Business Law Management, Financial Management

College tuition was partially financed through my personal employment.

EXPERIENCE

1/91–Present   CRESCENT DATA SYSTEMS, Richmond, VA
(Nationwide Chain of Retail Software and Hardware Items)

Sales Representative Retail Store - Sales of Hardware (Apple and PC) and Software; Microsoft Word, WordPerfect, Excel, Aldus Pagemaker, Microsoft Windows, Data Base IV and Lotus 1-2-3, among others. Responsibilities include consulting with customers on their needs, directing them to proper software, and demonstrating basics of software programs. Set up and install software together with hardware, sold as a package. Received seven awards for exceeding departmental sales targets.

1/89–1/91   THE CHOCOLATE BAR, Retail Store, Petersburg, VA
Assistant Manager In charge of the evening shift.

Responsible for a sales and stock force of 3 people. Maintain displays and promotion of boxed and loose candies and nuts. Keep perpetual inventory, actual ordering done by day staff. Train new stock people. Close out register and deposit evening revenue. Established new floor layout and significantly boosted average weekly sales totals.

4/87–1/89   HENRY PAYTON COMMUNITY CENTER, Lynchburg, VA
(Private Pool Club) Assistant Manager and Head of Maintenance

HOBBIES

Sports - Member of Y.M.C.A.

A superior personal endorsement leads the way.

MARY SMITH
45 Evansdale Drive
Anytown, STATE
(555)555-5555
E-mail: smith@network.com

"Perhaps the most effective data analyst who reported to me, and certainly the one with the best overall work ethic." (Ralph Summers, former Chief of Electronic Product Development)

GENERAL SUMMARY

Superior organizational, technical, and administrative skills. Over ten years business experience in a variety of settings. Continuous track record of systems analysis and programming applications that are maintained easily and quickly. Proven ability to utilize extensive knowledge of information systems.

EMPLOYMENT EXPERIENCE

HARON LEGAL CONSULTATION GROUP, Midland, Michigan
CORPORATION SYSTEM, New York, NY   **(2/92–Present)**

Programmer/Business Analyst
Electronic Product Development   **(10/96–present)**

Responsible for support of an integrated legal practice system which includes sub-systems for completion of corporate filings, jurisdictional searches, and document orders from federal, state, and local government agencies. Developed software configuration management procedures for a variety of systems, including ADW encyclopedia management, change control, problem management, and test strategies. Utilized Application Development Workbench (ADW) to maintain logical and physical data models, functional decompositions, data flow diagrams, structure charts and supporting documentation for PC-based system. Developed project standards and procedures, as well as trained others on the system. Plan and review deliverables of all life cycle phases; conduct modeling sessions with developmental and user teams before delivering presentations to executive management.

Programmer/Analyst
Data Administration and Application Development   **(7/95–10/96) Order Entry System/Mainframe Based**

Responsible for credit and billing requirements of an automated order entry system, analyzing existing functions and new system components as part of a development team. Defined and maintained corporate and enterprise information models; captured enterprise data, processes, and business functions to present graphically for purposes of analysis and verification. Interacted with Customer and Order Management Teams. Developed local procedures to control log-on id's, E-mail access and system authorization.

Programmer/Analyst
Financial and Administrative Systems   **(11/93–7/95)**

Activities centered on the maintenance of existing batch and on-line systems. Trained and supervised three junior programmers in the use of DPPX/DSX programs for purposes of distributing customer records and programs to branch offices throughout the country. Analyzed existing business systems to identify conversion, interface, and technical requirements. Assisted in the redesign and updating of company promotional mailing system including incorporation of structured programming for easy maintenance and generation of management reports.

Programmer
Applications Maintenance and Support   **(11/92–11/93)**

Coordinated program and CLIST maintenance, testing, and implementation for remote users. Trained and supervised two junior programmers to maintain and monitor data integrity between local and branch locations (twenty throughout the country).

Programmer Trainee
Application Support   **(2/92–11/92)**

Assisted in file maintenance for a distributed inquiry system. Provided technical support to remote users.

MARYLAND A&T STATE UNIVERSITY, Baltimore, MD   **1988–1991**
Clerk/Stenographer

Recorded and disseminated minutes, prepared materials and special reports for meetings, drafted reports, maintained files and membership lists, organized and drafted correspondence, and assisted staff with administrative duties.

HART-STAN FASHION STUDIO, Boston, MA   **1987–1988**
Advertising Assistant

Assisted advertising director in coordinating promotions and fashion show productions. This included public relations work, dealing with media representatives, supervising temporary personnel, periodic budget preparation and administrative procedures.

INTERNATIONAL EVENTS PRODUCTIONS, Boston, MA   **1986–1988**
Administrative Assistant

Worked with Director of Foreign Sales, assisting with administrative functions, dealing with such countries as Venezuela, England, and the Philippines, to procure broadcast rights. Maintained sales logs, prepared contracts, sales confirmations, and written/telex offerings; interacted with Accounting Department.

PROFESSIONAL TRAINING/EDUCATION
HAGERSTOWN TECHNICAL COMMUNITY COLLEGE, Hagerstown, MD
Associate Degree in Business Computer Programming - August 1986
President's List, 1986

RAVREDE COLLEGE, Ravrede, MD
Bachelor of Music in Vocal Performance - May 1977
Dean's List

Communitech Institute, Microsoft Windows, 3.0/3.1 - 1992 Systems Education Center, DB2: Concepts, DBA Workshop MVS: Concepts/JCL-1991 Net-Serve-Software, IE: Concepts, Data Modeling, S Programming & Design - 1990 Boston College, CICS Concepts and Facilities, DOS/VSE JCL - 1988–1989

COMPUTER KNOWLEDGE
Hardware: IBM 3090, 4381, PS/2, 8100/9370

Software: Windows, DB2, TSO/ISPF, QMF, MVS-ESA, WordPerfect, Fastback, CICS, VSAM, DPPX/DTMX/DSX, DOS/VSE, VOLLIE, ABEND-AID, DYNAM-D, PC-DOS, Lotus, Pascal, C, Excel, Paradox, Dbase IV

Case Tools: ADW/IEW

Languages: COBOL SQL, JCL

REFERENCES
Available upon request

An unusually detailed Professional Overview is justified by the applicant's significant experience and accomplishments. Note the inclusion of background information that helps the reader make sense of the educational record.

JOHN SMITH

45 Evansdale Drive
Anytown, STATE
(555)555-5555
E-mail: smith@network.com

OBJECTIVE

To obtain a position with a vital organization in the electronics field where my experience and background would result in improvements in profitability and operational efficiency.

PROFESSIONAL DEVELOPMENT

1990–present   PACE ELECTRONICS, Denver, Colorado
Design Engineer

The company primarily deals in government contracts for radar equipment, airplanes and other military equipment. My activities centered around the design of coaxial adapters, connectors, DC-blocks, detectors, terminations, chip attenuators, microwave switches on stripelines (LP & HP electronic oscillators, drivers, and PC-boards). This involved building prototypes for pre-production testing and evaluations; Designing tools and fixtures for assembly; creating, assembling, and testing procedures; successfully completed a course of study in measurements using the HP8510 Network Analyzer system at Corbett-Pollard in Colorado Springs, Colorado.

1983–1990   MANTEL SERVICES, La Junta, Colorado
Production & Engineering Department

My activities centered around the assembling, matching and the partial design of and testing of different waveguide, couplers, circulators, magic-T, flexible waveguides, transitions, terminations, antennas, ferrite isolators, switches, mixers and waffle iron filters. This included construction of, modeling, drawing, building of prototypes and testing wideband waveguides adapters and loop couplers; noise measurements and high power, impedance matching, as well as resolutions of mechanical problems.

1983   POWER TECH ELECTRONICS, Pueblo, Colorado

Designed, built and tested prototype for electronic circuits for smoke alarms, central heating and printed circuit boards.

1981–1983   INTERNATIONAL TRADE COMMUNICATIONS, London, England
Safety Instructor

1977–1981   COMMERCIAL COMMUNICATIONS INTERNATIONAL, Munich, Germany
Science Research/Senior Assistant Professor

Activities centered around the repair, calibration, and modernization and inspection of electronic and microwave equipment. This included graphics design and drawing; examining of semiconductor switches on coaxial lines for design and testing of variable attenuator on cylindric waveguides. Also responsible for conducting management workshops and teaching mathematics and physics, for about thirty students per semester.

EDUCATION

TECHNICAL UNIVERSITY OF MUNICH, Munich, Germany
The course of study is broadly based and moves progressively towards a selected area of specialization.
While a student, sent to different factories to learn various aspects of the profession. Degree awarded in 1968 in Electronics//Solid State Electronics (comparable to a Master Degree awarded by U.S. University - 6-1/2 years of study).

LANGUAGES

German, Chinese, and Spanish.

A list of appropriate adjectives sets the tone for a summary of superior
accomplishment as a manager and as a technican.

JOHN SMITH
45 Evansdale Drive
Anytown, STATE
(555)555-5555
E-mail: smith@network.com

Overview
Nineteen years of superior experience as an electrical design engineer with AIL division of PRESTON
CORPORATION.

*Organized * Motivated * Energetic * Attentive * Team-Oriented * Superior Manager*

1992–Present
Principal Electrical Engineer
Involved in modifying a prototype electronic data collection system to its production configuration. I
was responsible for building and testing new subassemblies as well as for the final system integration
and hardware/software acceptance testing. During this period I was also involved in generating Engineering Change Proposals for expanding the system's capability.

1988–1992
Senior Electrical Engineer
Responsible for the design and development of test equipment that tested subassemblies of an electronic
data collection system. This equipment contained analog and digital circuits, and consisted of board testers
and major subassembly test sets that performed manual and automatic testing. One test set employed an
HP desktop computer to perform end to end (IF input to digital data output) testing of an Encoder/Con-
troller. During this period I was also responsible for the design of system modifications and their incorporation into production systems. In addition I was involved in hardware test and hardware/software compatibility testing of the data collection system. I also supported testing of automatic computer controlled test
equipment which performed end-to-end testing of the data collection system.

1985–1988
Senior Electrical Engineer
Responsible for design and development of electronic assemblies that were part of a military airborne
data collection system. The units for which I was responsible were the following: IF and Analog Signal
Encoder, Analog Data Multiplexer, Digital Signal Encoder, and System Power Assemblies. During this
period I had one engineer and two engineering associates reporting directly to me. At various times I had
two consultant type engineers reporting to me on a part-time basis to provide expertise and designs in
specific areas. I was also involved in environmental testing and in system hardware/software compatibility testing.

1984–1985
Senior Electrical Engineer
Assigned to a program to modify an operational military airborne data collection system by replacing a
hardwired computer with a programmable one. I was involved in the generation of the computer specifications, especially the area of Input/Output interface requirements.

An unorthodox "numbering" approach makes this resume instantly memorable — and says
a lot about its author's passion for logical organization.

MARY SMITH

45 Evansdale Drive, Anytown, STATE, (555)555-5555, E-mail: smith@network.com

A tested, experienced professional specializing in the analysis and design of business software applications.

EDUCATION

North West Technical School, Yreka, CA
M.S.: Computer Science May 1983

California State University, San Francisco, CA
B.S.: Business Administration December 1979

EXPERIENCE

Mantec Company, Sacramento, CA
February 1991–Present   **Senior Analyst-Programmer**

Basic Function: The standardization of existing procedures and the development of computerized operational systems for an electronics manufacturing company. The computer facility consisted of a Digital P.D.P. 11,70. and Vax I I.'780 computers using I.A.S., V.M.S., Cobol. Fortran. Datrieve. and D.B.M.S. II.
Responsibilities: 1. Structured analysis and design of new and improved systems, including data base development and programming specifications. 2. Budgeting and cost-benefit analysis of all proposed systems. 3. Documentation of all systems. 4. Implementation of all systems 5. Supervise analyst-Programmer, as well as train and assign specific projects and tasks. 6. Liaison and interface between the Operations division and the M.I.S. steering committee.

Norcom Systems, Palo Alto, CA
April 1991–June 1993   **Systems Designer**

Basic Function: The development of C.I.C.S./Cobol application systems for a systems consulting group.
Responsibilities: 1. Design and code systems/programs. 2. Documentation of all systems.

Redwood Community Hospital Center, Red Bluff, CA
February 1990–January 1991   **Systems Analyst**

Basic Function: The management of the manual information systems for a hospital nursing department.
Responsibilities: 1. Structured analysis and design of new and improved systems and procedures. 2. Design of standard forms. 3. Documentation of all systems. 4. Implementation of all systems 5. Supervise data collection/processing staff. 6. Provide feasibility reporting for: A. Data processing B. Word processing C. Micro graphics.

Oakland Heights Service Corp., Oakland, CA
February 1978–January 1980   **Analyst-Manager**

Basic Function: The development of computerized business systems and the management of a computer facility for a liquor products company. The computer facility consisted of a Cromemco System Two computer, using C.D.O.S., COBOL and D base II.
Responsibilities: 1. Analysis of new and improved systems. 2. Design and code all systems including database development. 3. Budgeting and cost-benefit analysis of all proposed systems. 4. Documentation of all systems. 5. Implementation of all systems. 6. Supervise computer operator and programmer, as well as train and assign specific projects and tasks.

ACTIVITIES

Membership in the Association for Systems Management.

REFERENCES

Available upon request.

The use of specific computer languages and hardware and software names is important; it allows easy retrieval of the resume from computerized databases.

JOHN SMITH
45 Evansdale Drive
Anytown, STATE
(555)555-5555
E-mail: smith@network.com

SUMMARY OF QUALIFICATIONS:

Extensive experience in analyzing and solving complex problems involving programming. Consistent track record of systems analysis and programming applications that run efficiently and are maintained regularly. Demonstrated ability to apply sound knowledge of data processing concepts to diverse applications. Use of structural approach in programming.

LANGUAGES: COBOL, COBOL II, Adabas, CICS, Relia Cobol, IDMS-ADSO

HARDWARE: IBM 30XX, 43XX, 50XX

SOFTWARE: IDCAMS, VSAM, TSO/ISPF, ICCF, DOS/JCL, OS/JCL, (Utilities), AND SPECIAL MS/DOS, (Visible Analyst Package) flowcharts and dataflow PROGRAMS: diagrams, EASYTRIEVE

EMPLOYMENT:

4/96–present VEGA COMPUTER SERVICES, INC. Consultant
Installation of software package which handled segregation of securities in margin accounts and maintained bank loans at preset levels. Recommended modifications after analysis of on-site systems and assisted with the installation of required changes.

10/95–4/96 UNITED FINANCE CORPORATION Senior Systems Analyst
Analyzed current P & S System as a prelude to creating specifications for a new system.

7/93–10/95 FERRON & CO. Senior Programmer/Analyst
Project Leader for Credit Interest System which handled accounts totaling $20 - $30M. Wrote, tested and maintained the program which ran once a month crediting interest into accounts automatically. Maintained automated settlement system (CreditCo) with banks nationwide.

Systems Analyst in the Sweep System; worked directly with Vice-President dealing with 25 banks. totaling over $1M of yearly transactions. Determining account balances, Tefra/WRA taxes and redemption purchase orders. Wrote, tested and maintained CICS Program which adjusted messages for customers. Responsible for maintenance of the BTSI Margin System.

M.S. (Comp. Sc.) Pace University, New York. B. Com (Accty.) Calcutta University, India

HARDWARE: IBM 370, DEC VAX 11/780, UNIVAC.

SOFTWARE: PL/I, COBOL, IDMS, ORACLE, SQL, FORTRAN, OS/JCL, TSO, OS/MVS, VM/CMS, VAX/VMS, VSAM

A pertinent quote from a recent salary review gets the ball rolling.

MARY SMITH
45 Evansdale Drive
Anytown, STATE
(555)555-5555
E-mail: smith@network.com

PROFILE

Experienced, capable, innovative professional with record of effective participation in complex projects. Demonstrated competence in adhering to completion schedules (cited for "superior deadline management" in most recent salary review) and maintaining the highest quality of work.

PROFESSIONAL DEVELOPMENT

MULTITECH, Portland, Oregon   **Electrical Assembler**
B Activities focused on specific applications of government contracts with the United States Navy, i.e., Trident Submarines, Aegis Weapon systems, North Warning Radar System, etc. Worked closely within required specifications. Certified in 2000, 2000A and 454.

SILVESTRO ELECTRONICS, Lake Success, New York   **Electrical Assembler**
C Activities centered on subassemblies and assemblies for Naval systems, i.e., classified radar and defense systems, to be installed into higher assemblies at later dates.

Group Leader—Issued work assignments to assembly line personnel. Transferred finished equipment to inspection/test areas. Worked on first piece and prototype units and equipment from blueprints and schematics. Worked with Engineers on modifications of finished/in-process equipment and circuits. Projects involved the production of signal generators for the U.S. Air Force, working with high frequencies, Gigahertz and Megahertz microwaves.

A/B/C Assembler—Worked from blueprints, pictorials and schematics, reading and interpreting changes made by the Engineering team with regards to the modification of circuits. Worked with little or no supervision, both wiring and assembling units.

SPECIAL TRAINING

Electro Static Discharge, Safety Awareness/handling of chemicals and solvents. Knowledge of all types of tools.

MILITARY SERVICE

UNITED STATES ARMY—Worked with and maintained electronic equipment, communication units, field equipment and field services. Received several promotions.

References Available Upon Request

Relevant course work is displayed concisely and directly near the top of the resume. The applicant's limited work experience is bolstered by a powerful endorsement.

MARY SMITH
45 Evansdale Drive
Anytown, STATE
(555)555-5555
E-mail: smith@network.com

OBJECTIVE
A position in electrical engineering.

EDUCATION
Reedsport University, School of Engineering and Applied Science
Bachelor of Science May 1989
Major: Electrical Engineering
Projects: Designed and implemented a 16-bit D/A and A/D converter. Designed and computer simulated a digital filter.
Major course work: Digital Electronic Circuits • Signal and Systems Digital Signal Processors • Communication Systems Digital Logic • Electromagnetics Digital Circuits Lab • Simulation in Production Solid State Devices • Data Structures
Knowledge of Pascal.

EMPLOYMENT EXPERIENCE

Reedsport Astrophysics Laboratory   **Reedsport, Washington**
Programmer   *June '97–present*
- Optimized presentation of papers to be published with computer-aided programs.
- Consultant to administrative personnel in troubleshooting software packages.
- Praised for "scrupulous attention to detail and unflaggingly positive attitude" by Professor Ralph Hershey.

Reedsport University, Graduate School of Business   **Reedsport, Washington**
Computer Consultant   *June '96–Sept. '96*
- Assisted students with their programs.
- Assisted professor in conducting computer classes.

Reedsport University   **Reedsport, Washington**
Math Tutor   *June '86–Sept. '86*
- Tutored freshmen in first semester calculus.

HONORS AND AWARDS
United Federation of Teachers' Scholarship
The Washington State Library Award

INTERESTS
Bowling, chess and basketball.

References furnished upon request.

ELECTRICAL ENGINEER (COMMUNICATIONS EMPHASIS)

Here, the applicant, in search of his first job within the field, has opted to focus not on job titles but on skills and proficiencies within his area of specialty. The endorsement from a respected industry figure helps to bring the whole package together.

JOHN SMITH
45 Evansdale Drive
Anytown, STATE
(555)555-5555
E-mail: smith@network.com

"I wish I'd started my career with all of your energy."—Edward Norris, CEO, DVW Corporation

OBJECTIVE
To obtain a position as an electrical engineer with a firm specializing in electronic circuits and communications systems. Interested in the design, production and testing of components and systems.

EDUCATION
BACHELOR OF ENGINEERING (Electrical)
Frankfurt College of Engineering, Frankfurt, Germany

BACHELOR OF SCIENCE (Physics and Chemistry)
University of Bangkok, Bangkok, Thailand Received degree in June 1989

PROFESSIONAL QUALIFICATIONS
- Broad base of knowledge and experience in electrical engineering. Considerable laboratory and research experience in electronic circuits and communication systems as indicated by the following design projects:
- Design and testing of an upconverter for AM broadcast band reception
- Group leader of a senior design project for Spar Aerospace Ltd. (Montreal), involving the industrial simulation of a communications receiver for direct broadcast satellite.
- Design and testing of an infrared digital data communications system for use in polling routines in management conferences.

WORK EXPERIENCE
Summer 1991 INVENTORY SUPERVISOR Thailand Imports House, San Francisco, California

Overseeing inventory control for this clothing importer; also engaged in retail sales, constantly interacting with the customer.

PROFESSIONAL AFFILIATIONS
Member, Institute of Electrical and Electronic Engineers (IEEE).

EXTRACURRICULAR ACTIVITIES
Swimming, jogging, squash, travelling, wireless electronic communications.

COMPUTER LANGUAGES
Fortran and Assembler

LANGUAGES
English, Thai and German (good working knowledge).

PERSONAL DATA
Excellent Health...Willing to travel and/or relocate.

References available upon request.

The early lines of the resume outline the applicant's current academic status, and the supporting material provides a mixture of carefully selected accomplishments in both academic and professional settings.

JOHN SMITH
45 Evansdale Drive
Anytown, STATE
(555)555-5555
E-mail: smith@network.com

EDUCATION

Master's Physics, Boston College, December 1991. Part-time student; available for full-time employment. G.P.A.: 3.54

Master's Electrical Engineering   Boston College, June 1990 G.P.A.: 3.8

Bachelor's Electrical Engineering   Boston College, January 1989

EDUCATIONAL HIGHLIGHTS

Special Projects:

Tutorial program with Prof. Panacord, Chairman of Graduate Committee, studying systems of lenses; optics; fiber optics; conducting experiments with laser beams to study their gaussian behavior.

Control System Design for a fourth order system using the ACSL simulation package. Involved designing fourth order controllers; observers; tracking systems for exogenous variables.

Advanced Study (Optics) Power Systems Radiation & Optics

Semiconductor Devices Electromagnetics

Microwave Electronics Mathematical Physics I & II

Control Systems

COMPUTER SKILLS

Programmed in Pascal, and Assembly languages utilizing a mainframe VM/CMS computer system.

BOSTON COLLEGE, Boston, Massachusetts   Feb. 1991–Present
Senior Laboratory Instructor instructing a senior laboratory course for undergraduate students studying for Bachelor's in Electrical Engineering. This involves experiments in feedback analysis; design of compensators; measurements of UHF; impedance matching; power experiments. Work directly with Ph.D's; 40 students (1/2 semester), 20 students currently.

VOXTECH SERVICES, INC., Boston, Massachusetts   Jan. 1991–Present
Field Engineer monitoring excavation for construction sites such as Penn Central Control Center. This involves record keeping of blast vibrations, use of seismograph and air blast monitors. Installed strain gauges in order to monitor tunnel walls for the Transit Authority construction site at the Mass. Ave. tunnel. Periodically measure strain gauges for tension on wall, etc.

OTHER

Fluent French, both written and verbal skills.

References Available Upon Request

ELECTRICAL ENGINEER (FIELD SERVICE BACKGROUND)

The inclusion of the applicant's initial position with the current employer highlights her strong mechanical ability. Other positions listed reinforce detail orientation, a strong work ethic, and managerial ability.

MARY SMITH
45 Evansdale Drive
Anytown, STATE
(555)555-5555
E-mail: smith@network.com

Objective:
To obtain and secure a position that offers the opportunity for advancement in the field of Electrical Engineering.

Education:
AUGUSTA INSTITUTE OF TECHNOLOGY, Augusta, Maine
Degree awarded June 1987 BSEE (Electrical Engineering) NYIT Electrical Engineering Club

Courses Included:
Control Systems
Digital Control Systems
Electromagnetic Theory
Communication Theory
Microwave Engineering
Technical Writing

Work Experience:
5/94–Present PORTAL COMPUTER SERVICES, Portland, Maine
MCS Business Machine (a subsidiary). Hired as a Field Service Technician After completion of training on NP8000, 7000, and 2000 Copy Machines, AP400, 500, and 1100 series Word Processors, was assigned and became thoroughly familiar with the territory (boroughs of Queens and Nassau). With the use of the company car, serviced fax machines, copiers, word processors; also mechanical and electrical troubleshooting.

5/90–Present BANGOR STATE UNIVERSITY HOSPITAL, Bangor, Maine
Cashier—Parking Garage, Operated ticket printers and cash registers. In charge of troubleshooting equipment when necessary. Worked part-time while a student to help defray educational expenses.

8/93–5/94 MULTI FORM CORPORATION, Manchester, New Hampshire
Assembler/Troubleshooter (Part-time) In charge of assembling Temperature Monitoring System on printed circuit board. Once assembled, the units were put to rigorous test procedures to verify that they function according to manufacturer specifications. Should this test prove unsatisfactory then a system of backstepping is initiated to find the fault.

8/81–5/84 ROLAND PARK CORP., North Conway, New Hampshire
Shift Manager Responsible for daily inventory and monitored two armored car deposits. Supervised three men. Received gas shipments and made sure the process flowed smoothly.

Personal:
English Achievement Award, 1981
Knowledge of Pascal, Fortran, PL/M, Basic, Vax II
Fluent in Spanish and English
Knowledge of Word Processing
Member IEEE
Willing to relocate

References Available upon request.

The Objective section makes persuasive reference to both the applicant's breadth of experience and to the areas where her work will add value to the prospective employer's organization.

MARY SMITH
45 Evansdale Drive, Anytown, STATE/(555)555-5555/E-mail: smith@network.com

OBJECTIVE
A position as an Electrical Engineer where my broad range of experience (domestic and international) will lead to profitable solutions incorporating reduced costs, increased production, and lower defect rates.

EDUCATION
1985–1988
LOS ANGELES SCHOOL OF SCIENCES AND TECHNOLOGY, Los Angeles, California

A.A.S. Electrical Engineering Technology GPA: 3.3	Electrical Drafting
Digital Electronics I & II	Electronics
Electric Machine Theory	Electric Machine Lab
Electronic Controls	Projects Lab
Computer Systems I & II	Member, Sigma Epsilon Tau, Honor Society of Engineering Technologies

EXPERIENCE
1994–Present   Electrical Technician   STANHOPE GROUP, Glendale, California
This firm manufactures electrical devices and switches for residential and industrial purposes. My major activities center upon improving production capability, working with Engineering team. This involves working with blueprints and schematics, from basic elements to final products. Update existing equipment and redesign new production line machinery. This with robots, automatic controls, fiber optic sensors, AC/DC motors, 110/220 volts, and rewiring existing machinery.

1993–1994   Electrical Technician   SPERRY, CO., Milan, Italy
Focused on the maintenance and repair of electrical conveyors, elevators, electrical equipment. This included working with AC/DC electronic controls, relays, breaks, AC/DC motors, small compressors and air conditioners.

1989–1993   Assistant Electrical Engineer   MERCHANT LINES OF FRANCE, Marseilles, France
Primary responsibilities on board this 50,000 ton automated merchant ship included the maintenance of all electrical equipment (AC/DC), motors, motor controls, AC/DC generators, battery chargers, automatic elevator and crane maintenance (hydraulics) and power distribution.

1989   MADRID POWER PLANT, Madrid, Spain
Responsibilities included upkeep and maintenance of all electrical equipment at the coal generated power plant. This equipment included: generators, AC/DC motors and controls, cranes and conveyors at our substation (220 volts - 380 volts) and power distribution controls.

ADDITIONAL EDUCATION
MARINE COLLEGE, Marseilles, France (1989–1992)
Completed 40 credits towards B.E.E.

CITY TECHNICAL COLLEGE, Marseilles, France (1988–1989)
Completed one-year program in industrial electronics. (diploma issued)

REFERENCES
Available upon request.

ELECTRICAL ENGINEER (MANUFACTURING EMPHASIS)

Everything fits; even the references to non-technical jobs showcase
market awareness, number sense, and management ability.

JOHN SMITH
45 Evansdale Drive, Anytown, STATE/(555)555-5555/E-mail: smith@network.com

OBJECTIVE
To be a key individual on a computer management team assisting in the development and execution of policies and plans for profitable business growth.

Hardware: Installing and removing Disk Drives and Diskette Drives. The System Board (adding and removing SIMMS, upgrading CPU, installing the SCSI Chip, installing and upgrading Cache Jumper Settings, Software and System Error Messages, Connectors). Configuring the System (Basic and Advanced), System Security, Fixed Disk Physical Format (Setting the Interleave Factor). Installing and removing Motherboards. Upgrading PCs from 286 to 386 or 486. Installing and configuring Compact Disc. Troubleshooting Hardware and Software problems.

Software: MS-DOS, MS Word, Windows, Basic Language, C Language, Dbase, PC Tools, Norton, Lotus 123, Tango, Symphony, Autosketch.

EXPERIENCE
BIXOLD INC., Miami, Florida   **12/94–present**

Responsible for assisting with the development, coordination, manufacturing and production plans for electronic scale/control bar for a new product still to be introduced in the marketplace. Completed preliminary design reviews and production readiness reviews. As part of a team effort, developed the software, electronic engineering and hardware connections as part of manufacturing planning.

COMTECH, INC., Miami, Florida   **4/93–11/94**

Software/hardware maintenance on projects involving testing, evaluations, and troubleshooting IBM Microcomputer systems, insuring that all PCs were in operating condition for this advertising company in the business of making television commercials.

VIDEO COLLECTION INC., Tampa, Florida   **8/94–6/91**

Manager of this commercial video outlet servicing the Queens neighborhood. Analyzed accounts payable and receivable. Calculated payroll and commissions. Prepared company taxes and sales taxes. Made bank deposits and bank reconciliations. Processed and posted credit cards.

PARTS PLUS CO., Lake Worth, Florida   **4/81–8/84**

Manager for this wholesale auto parts outlet. Supervised a staff of 15 employees. Processed the daily and monthly inventory. Researched and analyzed the market for potential new products.

EDUCATION
HARDING CAREER SYSTEMS, Miami, Florida

Completed intensive training in Computer Programming and Operations, specializing in Microcomputers. Graduated with a G.P.A. of 3.7/4.0 two months ahead of schedule in 1993.

LEIPZIG INSTITUTE OF ECONOMICS, Leipzig, Germany Diploma received 1977

DRESDEN INSTITUTE OF TECHNOLOGY, COLLEGE OF MECHANICS, Dresden, Germany Diploma received 1976

References Available Upon Request

Passing reference to summer employment (in the Personal section) makes the most of past administrative and supervisory work— and highlights the applicant's ability to respond effectively when faced with an unfamiliar situation.

MARY SMITH

45 Evansdale Drive, Anytown, STATE/(555)555-5555/E-mail: smith@network.com

OBJECTIVE
Entry-level position in the field of electrical engineering with an opportunity for specialization in design.

EDUCATION
Jan. '97–Present
Corbett Institute, School of Engineering
B.E. Electrical Engineering - June 1998. GPA 3.84
Elective Courses: Computers, Microprocessors, Microprocessor Applications, Pulse and Digital Circuits, Digital Communication Systems, Modern Design and Development.

Sept. '94–May '96
Hastings College, San Francisco, California
Basic Pre-Engineering Curriculum - GPA 3.36

HONORS
Member of Tau Beta Pi, Member of Eta Kappa Nu, Daniel M. Wise Scholarship/Engineering President's List (Corbett Institute), Dean's List (Hastings College)

EXPERIENCE
June '91–Sept. 93
Prentor Electronics, Santa Clara, California
Responsible for answering phones, filing and making appointments for electrical contract work.

ACTIVITIES
Attended and aided the proceedings at the International Conference on Circuits and Computers, '96.

Member of the Institute of Electric and Electronics Engineers.

Member of the Corbett Engineering, election committee.

PERSONAL
Worked an average of 20 hours per week to help finance my college education; assumed supervisory duties when bakery owner was sidelined for two weeks by illness. Praised for "exemplary" managerial work during this period. Designed my own World Wide Web page (www.teruda/marysmith) and those of 15 other students. Proficient in 15 computer languages. Fluent in Spanish and Turkish. Willing to travel and relocate.

REFERENCES
Will be furnished upon request.

ELECTRICAL ENGINEER (RECENT LAYOFF)

Note the reference to temporary assignments with companies likely to
be recognized by a potential employer, and the information concerning
superior performance evaluations before the layoff.

MARY SMITH
45 Evansdale Drive
Anytown, STATE
(555)555-5555/E-mail: smith@network.com

QUALIFICATIONS
Engineering background includes experience with Optical Fiber Communications, Digital Control Systems, Microcomputer-Based Design, and Operational Amplifier Design.

COMPUTER SKILLS
Hardware: IBM-PC, 8085 and 8086 Microprocessors

Software: PSpice, WordPerfect, MathCAD, dbase III+ Languages: Pascal, Assembly, FORTRAN, BASIC Operating Systems: MS-DOS, IBM-VM/CMS, UNIX, VAX

EMPLOYMENT EXPERIENCE
8/96–present:	Temporary employment assignments through Bestway Technical Agency. Assignments have included: VeeJam Associates, Berrison Brothers, and Michael Creeto Group.
7/94–8/96:	ELECTRONICS SERVICES, Toronto, Canada Assistant Electrical Engineer Clients included one of Peru's largest electro-domestic companies - Moraveco. Activities included creating block diagrams for devices, using electrical wiring diagrams and physical drawings. Routinely called upon to assist Senior Engineers. Tested the accuracy of digital devices. (Received superior personnel evaluations; position eliminated due to company restructuring in 1996.)
5/94:	EDUCATION MIAMI DADE TECHNICAL INSTITUTE, Miami, Florida Bachelor of Science Degree. Major: Electrical Engineering, Coursework and special projects included: - Designed an equalizer and wrote a subroutine in Pascal to demonstrate Equalization concepts. - Designed a counting Analog-to-Digital converter. - Participated in research of general electromagnetic theory and corona discharge in transmission lines. - Designed a practical integrated circuit operational amplifier using commercially available components. - Designed an 8086-microprocessor-based function generator. - Designed an 8085-microprocessor–based fire alarm system for a six story building, using two 8212 latches, a decoder, an EPROM, and a RAM. - Conducted research on Optical Fibers to learn their advantages over their metal counterparts.

AFFILIATIONS
IEEE member Completed two years of R.O.T.C.

LANGUAGES
Fluent in Spanish, Proficient in French.

* Willing to Relocate.

The use of visually powerful "grabbers" to set off important chunks of text demonstrates the applicant's energy—and his willingness to work to make technical material accessible to others (a key skill in this position).

JOHN SMITH

45 Evansdale Drive, Anytown, STATE/(555)555-5555/E-mail: smith@network.com

PROFESSIONAL EXPERIENCE

LOS ANGELES FIRE DEPARTMENT, Los Angeles, California   *10/94–Present*

Electrical Equipment Safety Specialist ✓Conducted fire protection/prevention inspections of wide range of structures, equipment and materials, involving multiple dwellings, industrial facilities, commercial and institutional buildings. ✓Achieved high levels of safety at construction sites, pinpointing potentially dangerous conditions and using significant expertise in dealing with the Fire Code to enforce corrective procedures. ✓Ensured the proper installation and maintenance of boilers and other HVAC equipment, applying a meticulous approach during inspections. ✓Contributed to the proper set-up of new businesses, meeting with company principals and operations personnel to instruct on all Los Angeles City fire and safety standards to ensure initial compliance.

BETA ENGINEERS AND CONSTRUCTORS, San Diego, California   *10/90–10/94*

Associate Electrical Engineer ✓Supervised the design, installation and maintenance of intrusion alarms, public address, and lighting systems for approximately 100 New York City schools, as well as power distribution upgrades. ✓Supervised the successful design and installation of Class "G" Fire Alarm Systems and Electromagnetic Door Locking Systems at forty (40) N.Y.C. Public High Schools. ✓Participated in electrical power distribution upgrades and station rehabilitation projects for the N.Y.C. Transit Authority, performing tasks such as detailed field surveys, feeder/branch circuit design and lighting calculations and supervising installations. ✓Contributed to on-time and minimally disruptive rehabilitation of 21 New York City police precincts, participating in design, cost estimating, shop drawing review, scheduling and monitoring of mechanical, electrical and plumbing trades, as well as the phasing for asbestos removal projects. Provided electrical, mechanical and construction expertise to various rehabilitation and renovation projects.

LOS ANGELES CITY BOARD OF EDUCATION, Los Angeles, California   *1988–1990*

Assistant Electrical Engineer ✓Supervised asbestos abatement programs to ensure compliance with DEP and EPA requirements. Prepared specifications for contract bids and cost estimates.

H.W. MOORE TECHNICAL SERVICE, Los Angeles, California   *1986–1988*

Electronics Component Engineer ✓Designed electronic components testing procedures in accordance with military specifications, and provided direction to technicians.

EDUCATION

HASTINGS INSTITUTE Degree: Bachelor of Science, Electrical Engineering, 1985

The excerpt from a written letter of appreciation is particularly effective here.

JOHN SMITH
45 Evansdale Drive
Anytown, STATE
(555)555-5555
E-mail: smith@network.com

EDUCATION
City College of Minneapolis-School of Engineering, B.E. Major - Electrical Engineering
City College of Minneapolis, M.E. (Anticipated) Major - Electrical Engineering

EXPERIENCE
1985–Present BAINBRIDGE ENGINEERING ASSOCIATES, Minneapolis, MN
Minneapolis International Airport

Position: Electronics Engineer Airway Facilities Division Electronics Engineering Branch Communications/Interfacility Section.

My responsibilities include designing devices such as antenna supports, layout junction boxes, mounting for control panels, and impedance matching devices for transmitters and other equipment; assembling engineering data; reviewing sources of information and technical literature; preparing reports for review by higher grade engineers; installing, modifying, testing, adjusting and tuning electronic facilities; directing the termination of cables and lines and the connection of antennas to transmitting and receiving equipment; and designing circuits.

Completed the following projects: 2/86-ICSS; 5/86-BUEC (2 projects); 8/86-RCAG (2 projects);10/86-Design a Monitor Panel for Non-directional Beacons; 10/86-RML-6-. Received letter of appreciation from the management and department citing my "superior, detail-oriented work" as an "asset to the department and the organization as a whole."

1984–1985 BELA ENGINEERING CORPORATION, Chicago, IL
Position: Electrical Engineer - Trainee

My responsibilities included involvement with design of combined power and light distribution systems.

PROFESSIONAL AFFILIATION
Member - Institute of Electrical and Electronics Engineers

DATA
Geographic preference - California

REFERENCES
Furnished on request.

The highlights of a strong academic career are quite capably handled here— and note the inclusion of (only) relevant professional experience from the construction industry.

JOHN SMITH
45 Evansdale Drive
Anytown, STATE
(555)555-5555/E-mail: smith@network.com

EDUCATION:
ALLENTOWN UNIVERSITY, Allentown, Pennsylvania
B.S. in Electrical Engineering, expected May 1998
Grade point average: 3.61
   Electrical Engineering Coursework Electrical Circuits I/II
   Electromagnetic Fields Signals and Transforms
   Feedback System Principles Transmission Lines and Waves
   Five Semesters E.E Lab Solid State Devices and Electrical Machinery I
   Circuits I/II/III
   Principles of Communication Systems
   Computer Science Coursework
   Computer Programming I
   Switching Circuits and Digital Systems
   Introduction to Computer Architecture

AWARDS:
   College: Dean's List (seven semesters)
   Board of Trustees' Full Scholarship Recipient
   Senior Honor Student.
   High School: Valedictorian, N.Y.S.
   Regents Scholarship Winner
   Queens College President's Award for Achievement
   Minerva Award -(N.Y. Public Library)

EXPERIENCE:
1994–present   *J.P. KILKENNY & CO., INC., (Electrical Const.) Altoona, Pennsylvania*
Updated street lighting as built drawings for various traffic jobs. Designed and produced various street lighting machinery drawings for field use. Assisted in drafting of schematic wiring diagrams for Riverbend Avenue Bridge. Edited and maintained filing system on IBM PC using Lotus 1-2-3 and dBASE III software packages. Maintained and prepared progress reports for numerous company projects.

ACTIVITIES:
Intramural football, basketball, hockey, softball; Working knowledge of BASIC, PASCAL, SPICE and machine language.

PERSONAL:
U.S. Citizen

REFERENCES:
Promptly provided upon request

Relevant skills and accomplishments are set off in concise bullets
here, a particularly effective technique.

JOHN SMITH
45 Evansdale Driv
Anytown, STATE
(555)555-5555/E-mail: smith@network.com

OBJECTIVE
To obtain a position as Electronics Technician with a company offering growth opportunities, using my extensive experience in the field.

EMPLOYMENT BACKGROUND

2/89–Present   ASSOCIATIVE COMMUNICATION, Denver, Colorado
Electronics Technician
RESPONSIBILITIES: Duties include testing and troubleshooting power supplies and also analog and digital PC boards to the component level.

2/88–2/89   VISITECH COMPANY, INC., Pueblo, Colorado
Electronics Technician
RESPONSIBILITIES: Duties involved work as a Laboratory Technician for an electronics manufacturing firm *Experienced with building breadboards and test fixtures *In charge of testing and troubleshooting prototypes *Liaison with professional Engineering staff for support services

4/83–11/87   BURLOW HESTON, INC., Denver, Colorado
Electronics Technician
RESPONSIBILITIES: Varied duties for large electronics firm on Long Island *Testing and troubleshooting of power supplies and relay circuits *Troubleshooting to the PC Board and chip level *Knowledgeable about complex digital and analog circuitries *Experienced with binary notation and solid state semiconductor devices *Test equipment: oscilloscopes, frequency counters, signal generators, bridges, generators, power meters, RMS voltmeters *Conducted acceptance tests.

1981–1983   COLLECTED SECURITIES CORPORATION, Denver, Colorado
Carrier

1978–1981   DENVER TIMES, Denver, Colorado
Deliverer and Subscription Manager

1974–1978   MESA MESSENGER SERVICE, Denver, Colorado
Carrier and Filing Clerk

EDUCATION:

1988–Present   PUEBLO COMMUNITY COLLEGE, Pueblo, Colorado
A.S. Program, Computer Sciences

     ELECTRONICS TECHNICAL SCHOOL, Denver, Colorado
A.A.S. Electronics Engineering Technology, 1983

MILITARY BACKGROUND:
U.S. Army Reserves, Denver, Colorado (1983-Present)

INTERESTS:
Jogging, yoga, reading, travel.

References upon request

The applicant's superior academic record is summarized concisely—and accorded a high-visibility spot near the head of the resume.

JOHN SMITH
45 Evansdale Drive, Anytown, STATE/(555)555-5555/E-mail: smith@network.com

CAREER OBJECTIVE
Seeking to utilize my practical knowledge and experience in an Electronic/Computer Technician position to deliver profitable solutions to technical challenges.

EDUCATIONAL BACKGROUND
Currently enrolled in San Francisco Institute of Computer Sciences Electromechanical Computer Technology program.
Harding Technical Careers, San Francisco, California, Associates Degree in Electronic Circuits and Systems. Class Rank: 2/100 G.P.A.: 3.7/4.0. Selected to Dean's List.
Corbett University, Berkeley, California, Associates Degree in Science, Major: Electronic Data Processing, Worked part-time/full-time during schooling, paying 100% of educational expenses.

EXPERIENCE
1994–Present
TOSH CORPORATION
Richmond, California   Electronic Technician
Responsibilities include:
- Testing and troubleshooting electronic equipment (down to component level) relating to radar defense system on SYTROM project.
- Maintenance and repair of system level multiband receivers, RF and analog equipment, receiver processors and subassemblies.
- Testing-troubleshooting-repairing and utilizing Eaton built digital test equipment.
- Working directly with project engineers on system design revisions and modifications.
- Participating in controlled testing operations, using computer interfaced environmental chambers.
- Providing accurate, detailed documentation on all work performed.
- Using state-of-the-art equipment, i.e. network analyzers, logic analyzers, oscilloscopes, multimeters.
- Abiding Secret Clearance classification regarding confidential information.

1990–1994
QUINCY MEMORIAL MEDICAL CENTER
Piedmont, California Department of Radiology   Clerical Assistant
Responsibilities included:
- Preparing patients' x-rays for interpretation by physicians.
- Maintaining departmental billing records system.

AFFILIATIONS: Institute of Electrical and Electronic Engineers

REFERENCES: Provided promptly upon request.

An unforgettable headline makes the applicant's significant technical accomplishments instantly accessible.

JOHN SMITH
45 Evansdale Drive, Anytown, STATE/(555)555-5555/E-mail: smith@network.com

If You Saw the Space Shuttle Land Safely After Its Last Mission, You Were Watching My Handiwork.

OVERVIEW
Extensive background in repair and fatigue testing of structure and prototype work.

EXPERIENCE
11/95–Present BURTON SPACE SCIENCES TECH., Decatur, Illinois
5/95–11/95 BDA GROUP, Peoria, Illinois

Fatigue Testing: Performing layout and setup of test fixtures and structural repairs; worked on E2C fatigue testing (Structural Mechanic). Performed layout work on A6F and F14A for fatigue testing. Activities also included some hydraulic work.

Product Development (Prototypes): Constructing detail parts and subassemblies from blueprints and engineers' specifications. Reworking parts and laying out prototypes. Major reworking on shuttle training aircraft reverse thruster.

Worked on LAVI Wing Project, fatigue testing and composite testing; made ribcaps and miniribs for redesigned space shuttle wing assemblies; made DMR parts from raw stock and according to DMR Disposition.

Aircraft Projects: F14A, F14D, E2C, C2A, X29, A6E, A6F, Ranger.

MILITARY
1990–1994 UNITED STATES ARMY

Aircraft Maintenance Specialist Performed launch and recovery maintenance; aircraft structural integrity testing and engine and corrosion control inspections. Responsible for aircraft refueling/servicing (fuel, oil, liquid oxygen and nitrogen).

Received extensive cross utilization training in hydraulic environmental control system, electrical and engine areas. Responsible for detailed, accurate documentation and technical manuals. Trained and supervised subordinate personnel.

EDUCATIONAL TRAINING
BURTON TRAINING SEMINARS. Huck Blinds Rivets Installation, Burton Bolts Installation, Jo Bolts Installation, H-loks Installation, Grumman Fastner Hole Preparation, Fiberglass Composites Lay-up, Composite Trim Operations, Composilok II Installation, Composite Hole Preparation.

MILITARY TRAINING
Introduction to Aircraft Maintenance, 500 hours Aircraft Maintenance Course, Sheppard Technical Training Course, 6 hours F-15 Organizational Maintenance Manuals, 248 hours F-15 CAM-T Phase 3, Integrated Aircraft Systems.

REFERENCES
Available upon request

FIELD ENGINEER

Note the detailed and specific account of significant savings realized that appears near the end of the Esway Data summary.

JOHN SMITH
45 Evansdale Drive, Anytown, STATE/(555)555-5555/E-mail: smith@network.com

QUALIFICATIONS
Strong technical ability in troubleshooting and repairing computer systems including peripherals and mechanical devices.

Heavily involved in critical customer responsibility. As an on-call field engineer have developed excellent relations with customers by showing genuine concern for their problems and by keeping downtime to an absolute minimum.

EDUCATION
Certificate - Computer Science and Electronics, Benton Technical School (9/86–3/88)

Certified Electronic Technician (1987)

EMPLOYMENT
Esway Data Corporation, Miami, Florida, 4/89–Present
FIELD ENGINEER - Provide on-site maintenance of many types of computer equipment including minicomputers, disk drives, magnetic tapes, printers and communications devices. Perform preventive maintenance and diagnosis/fault isolation at the component and board level. Responsible for repairing all mechanical devices and peripherals.

Personally installed several small minicomputer systems. Was lead over two to three other field engineers on larger installations. Involved with installing several mainframes. Frequently relocate systems. As Site Engineer for Rona Aircraft (6/89–6/90) drastically reduced response time and downtime. Analyzed inventory needs and reduced significantly, saving Esway Data at least $15,000.

For the Midwest branch, responsible for monitoring factory change orders and ensuring changes are made at all customer sites. Assist Inventory Control Manager in keeping branch inventory and site inventories at proper levels

Mainframe Electronics, Macon, Georgia, 7/78–4/79
FIELD ENGINEER - Maintained financial processing equipment on-site, including check processors and C.O.M. units. Worked closely with customers to solve their problems.

Hamill Services Corporation, Macon, Georgia, 3/78–7/78
ELECTRONICS TECHNICIAN - Performed bench repair and calibration of all varieties of mechanical and electronic UHF/VHF tuners used in televisions. Handled inventory control duties.

A downsizing campaign led to a job outside the applicant's initial area of interest—but this resume's Objective section makes lemonade out of lemons!

MARY SMITH
45 Evansdale Drive
Anytown, STATE
(555)555-5555
E-mail: smith@network.com

OBJECTIVE

An entry-level or trainee position as a Field Service Technician/ Representative with Acme Technologies utilizing my substantial skill base and hands-on experience in electronics and electro-mechanical operations, as well as my demonstrated skills in customer service.

EXPERIENCE

6/95–Present
CARPET-RITE CARPET AND FLOORING, Dallas, Texas
Position: SALES/INSTALLATION Extensive customer contact. Disseminate information, make suggestions and close sales; frequently go to customer site to insure satisfactory installation procedures are followed.

9/93–6/95
RIO GRANDE DATA, El Paso, Texas
Position: Field Service Representative - El Paso Territory Responsible for installation, maintenance and repair of telecommunications equipment; working with varied clientele, on a contract basis; following up referrals and establishing new accounts; coordinating activities with Sales Representatives; establishing maintenance schedule and insuring follow through on all equipment; updating files; maintaining service diary; attend monthly meeting on maintenance updates and new equipment; knowledge of RS 232. Types of equipment serviced included: aeronautical radio equipment in airports as well as installation of satellite disks and accessories, such as: Extel - ASR and RO., Teletype - Model - 40,43,28,32,33,35 and 15 all varieties - ASR/BSR/RO, Zintec - CRT Model 50 and 90, Seimens - Model T-1000 and PT 80, SR/RO/CRT Models, G.E. - Model 30, 200, 300, 1200, and 1232, Modems, Datasets, DAA (all types, DOD, Tone, Loop, etc.). Test equipment: Multimeter, DB Meter, Scope, Headset, Breakout Box.

1990–93
PROCAMP COMMUNICATION, Ft. Worth, Texas
Position: Customer Service Option Technician. Responsible for installation of options according to customer specifications on Electronic and Electromechanical Telecommunications Equipment; testing equipment on-line. In addition, experience in the rebuilding and refurbishing of varied computer terminals (Teletype, Zintec and G.E., all models). Company downsizing campaign led to elimination of position.

1988–89
MECHANICAL PROFESSIONALS, Ft. Worth, Texas
Position: Machinist, Machining soft metals; silver soldering.

EDUCATION

1982–84
FT. WORTH COMMUNITY COLLEGE (CUNY), Ft. Worth, Texas
Accumulated credits in Bachelor of Science Degree. Major: Mechanical Engineering
Training: CASING TECHNICAL SCHOOL, Mineral Wells, Texas Received Certificate, Basic Electronics
PROCAMP COMMUNICATION Courses in Electronic Repairs

REFERENCES AVAILABLE UPON REQUEST

GRAPHICS SPECIALIST

A well-organized summary of accomplishments in four results-centered
"acts" and an "epilogue" outlining software knowledge.

MARY SMITH
45 Evansdale Drive
Anytown, STATE
(555)555-5555/E-mail: smith@network.com

A skilled, resourceful professional with deep experience in a wide variety of printing and graphics technologies.

EXPERIENCE
11/90–Present
A.T. Graphics Associates, Boston, MA
    Coordinated and supervised removal and new installation of prepress/high technology equipment (cameras, digital image manipulation software, scanners, processors, platemaking equipment, step and repeat machines, etc.). Prepared schedules for various vendors, such as trucking and riggers. Responsible for maintaining smooth work flow. Drafted layouts and floor plans for complete preprep area consisting of dark rooms, contact rooms, camera and plate department, stripping area, etc. Supervised rebuilding of old equipment. Worked extensively in the field with customers.

1988–1990
Apex Design Group, Boston, MA
    Production coordinator—Large reproducing/graphics house, with following departments: Reproduction, Silk–Screening, Bindery, Pressroom, Camera/Platemaking/Stripping, Art and Typesetting, Video Graphics. Oversaw all phases of work, from conception to completion. Wrote job tickets, sized up mechanicals and laid out work to appropriate departments; scheduled deliveries. Responsible for attending to customers' needs and following up on all work to ensure timely completion. Maintained inventory of papers and supplies.

1985–1988
Barnett Communications, Inc., Lowell, MA
    Production supervisor—Responsible for billing (over two million dollars in sales) and preparing estimates from account executives' quotations. Handled magazine work for advertising agencies. Interacted with outside vendors, dealt with sales representatives; extensive customer contact. Wrote up B/W and 4/C job tickets, sized up mechanicals and distributed work out to shop; followed up on all work.

1983–1985
Bergen Creative Group, Springfield, MA
    Production supervisor—Purchased plates, chemicals, paper, miscellaneous supplies. Estimated from 8-1/2×11 to 18×25 presses. Wrote up job tickets and dealt with customers and outside vendors.

    Commitment to ongoing personal development has yielded exceptional skills in: PageMaker, Quark, Photoshop, and all Microsoft Office applications.

Instead of a single extended paragraph, this applicant has broken the Summary of Qualifications into seven relatively brief, stand-alone sentence that provide instant insight into his experience and qualifications.

JOHN SMITH

45 Evansdale Drive, Anytown, STATE/(555)555-5555/E-mail: smith@network.com

OBJECTIVE

To obtain a position as a Prototype Wireman or Electronic Technician where my extensive experience will be mutually rewarding.

SUMMARY OF QUALIFICATIONS

In charge of Electronic Department operation, including supervision and training of line technicians.

Assembly and troubleshooting of prototype mechanisms; working in conjunction with Engineer, developing effective apparatuses.

Traveling to other states demonstrating, installing, servicing and training customers in proper and effective use of equipment; attended several Electronics Seminars.

Knowledgeable and experienced in installation, assembly, testing and wiring of electronic equipment, components, PC boards, satellite antenna.

Adept in proper use of specialized tools and test equipment.

Preparing Department Report detailing status of machines; ordering and reporting on parts inventory.

Organizing and ensuring completion of workflow by deadlines during extended summer production rushes.

PROFESSIONAL EXPERIENCE

1986–1997
BURBANK COMMUNITY COLLEGE, Burbank, California, Head Electronic Technician
1984–1986
Worked for a variety of companies, such as Find-A-Job Co., Harriet Bond Corp., Banks Electronic Systems, First Print Electronic Corp.
1983
HORIZON CORPORATION, San Diego, California, Sample Harness Maker
1977–1982
ELDEN TECHNOLOGIES, Pasadena, California, Prototype Wireman

EDUCATION

COMMUNICATIONS INDUSTRIES INSTITUTE, Los Angeles, California. Completed 825-hour Radio & Television Electronics Programs.

RELEVANT DATA

Previously licensed by NASA; Enjoy troubleshooting component systems; Willing to relocate.

INSTALLER/MAINTENANCE TECHNICIAN

A detailed but thoroughly results-oriented summary of technical capabilities.

JOHN SMITH

45 Evansdale Drive, Anytown, STATE/(555)555-5555/E-mail: smith@network.com

PROFESSIONAL EXPERIENCE

12/89–Present

GLOBAL COMMUNICATIONS, INC.

Installer/CSO/Maintenance Technician, Technical Support Engineer IISL1 VL, LE, VLE, XL, M, MS, N, XN, XT, SL-10 and SL-100 Systems. Installer: Lead man in charge of a crew from 2 to 10 men in the installation of an Integrated/Voice Data PBX from 200-2,000; lines erected MDF and IDF frames for Call terminations, to include cross connecting on both frames; mounting common and peripheral equipment; cabinets installation of Telco lines (MCI, SBS, Tie, DID, DOD and WATS). Installation of Call Detail Recorder and Paging Systems. System grounding and maintenance repair techniques.

Repair of 2 wire, 4 wire E&M, DID, COT (Ground and Loop Start Circuits). Repair of Central Processing Units, Data Base for Automatic Route Selection, Network Automated Route Selection and Basic Automatic Route Selection. Battery maintenance and peripheral equipment maintenance. Interface with various vendors MCI, SBS, NYNEX Telephone Co. tie lines, off-premise extensions; RS 232 interface with personal computers and data set memory and system reconfiguration of attendant consoles.

Designated maintenance and repair technician for New York area SL-10 packet switch (located at Federal Reserve Bank). Duties included upgrading data parts, testing modem pools and overall preventive maintenance.

Customer Service Order Tech: Responsibilities include the handling of all customer moves and modifications in data base and system hardware, peripheral and common equipment data set modems. Removal and addition of private lines COT, DID and Tie lines. Relocation of telephone sets, attendant consoles, bells, chimes and intermediate distribution frames.

1994

SL-100: Maintenance and customer commissioning experience (Meridien Trust Company). Resident TSE on site responsible for business set integrity, trunk maintenance and overall switch maintenance.

1996

Transnational Communications: Maintenance TSE responsibilities to maintain switch integrity and troubleshoot trunk and line malfunctions, i.e., voice, CP suicides and inoperable IBM lines.

1998

Philadelphia Hospital, Pennsylvania State College Medical Center: Maintenance and cut-over Commissions. Installation of telephone lines for business sets, testing switch apparatus for proper specifications. Install fiber optic lines for trunking application interfacing with compatible carriers for customer usage, integrating SL-1 systems with SL-100 translations, implementing network configuration for ultimate cost-effecting usage. Responsible for maintaining the high standard of SL-100 application.

Training: SL-100-DMS Commissioning/Maintenance, SL-100-DMS Translations, SL1 - Installation and Repair SL1 - N, XN, Maintenance SL1 - Data Base Management SL10 - Installation and Maintenance.

3/87–7/89
KORAL EQUIPMENT LINE, Richmond, VA Computer Specialists Employed as one of four Computer Specialist to interface with mainframe computers on a two-week rotation covering three shifts. Responsibilities included the following: Maintaining the B4700 Real Time System for users, data input and output, executing batch programs for data base daily updates, and creating custom forms for containerized items arriving from major European ports. In addition to these duties, utilizing other systems, such as the B3800, to update accounting files and to ensure that the Accounts Receivable and Trial Balance are without error. Other responsibilities required preserving the integrity of all the operating systems (PDP DEC 1140, Xerox 1200, and the RCA Mohawk Global Data Voice) to their most efficient level.

2/79–3/87
COOLIDGE SAVINGS AND TRUST, Pittsburgh, PA. Senior Computer Operator Employed as a coordinating and scheduling control clerk with the following responsibilities: Organization of check for computer processing and to distribute results for finalization and verification. Results: Schedules were met with a large amount of monies cleared for bank credit.

1980
Transferred to Computer Dept. as a Jr. Operator, primary duties to assist Sr. Operator with computer functions.

1982
Advanced to Sr. Operator, responsibilities include overall supervision of input/output. Control of work flow, repair minor computer malfunctions.

1985
Advance to Console Operator, primary duties: Maintain and supervise B4700 computer functions. Executing master control programs, patching records in disk directory. Scheduling programs for user processing and creating magnetic tapes for IBM 3700 input. Other duties, require testing new programs for validity and debugging techniques.

10/76–10/78
UNITED STATES ARMY Honorable Discharge: Not in Reserve

9/75–10/76
INTERNATIONAL SERVICES CORPORATION, New York, New York. Advertising Correspondence Assistant

ACADEMIC
9/81–6/83
PITTSBURGH COMMUNITY COLLEGE, AAS Degree, Major: Accounting Minor: Data Processing

An excellent array of academic credentials and real-world experience makes
a compelling case for an entry-level position. The decision to grant the Dean's
list achievement a single line is unconventional, but effective.

JOHN SMITH
45 Evansdale Drive
Anytown, STATE
(555)555-5555
E-mail: smith@network.com

OBJECTIVE
An entry level position that would enable me to deliver solutions in the telecommunications field for ABC Engineering.

EDUCATION
BOSTON COLLEGE, Boston, MA
Completed a Bachelor of Science degree in Engineering Technology, December 1996. Specialized in telecommunications and electronics. Member of IEEE. Attended Houston Community College during senior year of high school. Made Dean's Honor Roll.

SIGNIFICANT COURSES
Telecommunications Industry • Quality Assurance • Telecommunications Mgmt. • Data Communications • Telecomre Lab • AC & DC Circuits Control Systems • Microprocessors • Technical Writing • Electronic Devices Digital Instrumentation & Control • Applications of Elect • Managerial Communications Digital Telephony • Included hands-on lab time

WORK EXPERIENCE
1996 FOTOTRONICS, Boston, MA
Salesman . Demonstrated computer capabilities and instructed prospective customers on the use of various software. Handled repair log and complaints.

1995–1996 PARAGON CABLE, Boston, MA
Trained for telemarketing and outside sales representative. Installed converter boxes, hooked up VCR's, sold cable subscriptions and notified nonpaying customers.

1992–1996 Construction Boston, MA and Manchester, NH
Supervised Laborers BURTON CONSTRUCTION. Part-time (summers, vacations, weekends).

1979–1982 UNICOM, Boston, MA
Positions Held: sacker, cashier, and night-time assistant manager.

REFERENCES AVAILABLE UPON REQUEST

Here's a resume that gets right to the point. The concise, direct format goes a long way toward helping the entry-level aspirant to set herself apart from the crowd, and the summary of accomplishments within a temporary position is extremely effective..

MARY SMITH
45 Evansdale Drive
Anytown, STATE
(555)555-5555
E-mail: smith@network.com

OBJECTIVE
Seeking an entry-level programming position with an established firm to utilize and enhance my experience in the field of data processing.

EDUCATION
DANA HALL UNIVERSITY Bachelor of Science Degree: May 1997 Major: Computer Science Minor: Business GPA: 3.25

PROFESSIONAL EXPERIENCE
November 1997 to Present
GIBBS COMPUTER SCIENCES INSTITUTE, Baltimore, MD

Junior Programmer/Administrative Assistant
Prepared BMS maps on CA-FLEXISCREEN, helped prepare JCL for batch programs, as well as various other duties as part of a data conversion team that converted a UNISYS environment to an IBM environment for a COBOL/CICS application system. Currently employed as a full-time temporary employee while attending night school to complete my degree requirements.

REFERENCES
Available Upon Request. Travel Acceptable, As Is Eventual Relocation.

Compelling, relevant success stories—documents produced under tight deadlines, the aversion of a potentially catastrophic data loss—add interest and impact to this resume.

MARY SMITH
45 Evansdale Drive
Anytown, STATE
(555)555-5555
E-mail: smith@network.com

OBJECTIVE

A challenging position as a Local Area Network Manager for ABC Systems.

EXPERIENCE

DATA TRANSFER, INC., San Francisco, CA

Data Base Programmer/LAN specialist   **9/92–present**

Wrote, tested, and debugged programs. Wrote effective user guide for multiple-user data entry system under tight deadline. Customized user reports and data entry. Trained new users and maintained system.

GLENDALE TRANS CONTINENTAL AIRLINES, Glendale, CA

Computer Operator/Internship   **8/90–8/92**

System used PC (IBM Compatible)with LAN, NCR, OS, DOS, Multiplan, Quattro Pro, Q&A, CICS, and NetWare NetWork. Created and modified cargo and baggage claims data base files. Installed hardware and software. Created graphics and ran claims reports. Provided user training on a functional level. Supported LAN: Handled troubleshooting duties, and established automated backup procedures for systems that were demonstrated effective when a power outage shut down facility.

EDUCATION

HASTINGS COLLEGE, San Francisco, CA

B.A. Computer Science January 1992

Honors: Two Academic Achievement Certificates

Computer Languages: Pascal, Assembler, C, Basic, and Prolog

Software: All versions of Windows, Microsoft Office, Novell Advanced Netware 286/386 V2.15 Network, DOS, MS-DOS, Q&A, Excel, Lotus 1-2-3, PC Paintbrush IV Plus, PC Globe, PC Write, Turbo C, Turbo Pascal, Quattro Pro, and dbase III+.

CEDRONE BUILDING AND MAINTENANCE SERVICES, Los Angeles, CA

Maintenance 7/82–2/85

Repaired equipment; responsible for inventory. Supplied staff with appropriate equipment. Trained personnel in equipment operation and maintenance.

Enjoy reading, swimming and writing programs.

References available upon request.

Who says you have to list your work experiences in order? A recent stint as a driver after an industry downturn is de-emphasized by means of a brief chronological summary at the conclusion of the resume.

JOHN SMITH
45 Evansdale Drive
Anytown, STATE
(555)555-5555/E-mail: smith@network.com

Experience Summary
Mechanical Engineer with over eighteen (18) years of experience in engineering design and construction of various utility projects, such as fossil and nuclear-fueled electric generating station.

Professional Highlights
Ethen Services Corp., Engineer, Applied Mechanics Department

A mechanical engineer assigned to the Applied Mechanics Department. Recent assignment included site engineering position at the IVA Watts Bar Units 1 & 2 and Chicago Utilities, Aurora Electric Station Units 1 & 2. Responsible for design verifications, spec conformance checks and evaluations, modification of structural supports, and field construction interface. Reviewed for constructability of designs, generated the necessary design changes, and authorized and documented all required field modifications. Also performed pipe support design modifications as required in the field. In addition, provided cable tray evaluation for Civil/ Structural Department. Performed detailed computer analyses of relevant data.

Served as construction engineer. Duties included inspection of equipment and hardware, preparation of construction travelers and work packages, and Design Change Authorization. Also acted as liaison between other engineering disciplines to reconcile nonconformances and construction deficiencies, as well as to provide necessary documentation.

Former assignments included the stress analysis of piping systems used in prototype ships and submarines under contract with the Carter Atomic Laboratory. Primary responsibilities include the stress analysis of piping systems and support structures for nuclear and fossil plants in accordance with the ASME Section III and ANSI B31.1 Codes. Other responsibilities included the testing and verification of new developments and implementation of the in-house PIPESTRESS 2010 program for accuracy and code compliances.

Have a thorough knowledge of ASME Sect Ill, ANSI B31.1, and AISC codes. Have a solid working experience with PIPESTRESS 2010 and NUPIPE II piping programs, STRUDL finite element program and BASEPLATE program.

Ellison & Taper Engineering Corporation, Associate Engineer- Engineering Mechanics Division

Associate Engineer in Pipe Stress Analysis Group. Performed stress analysis of Safety Class 2 & 3 piping and related components for nuclear power plants. Also reviewed the feasibility of support designs and locations. Duties included the preparation of isometrics, stress analysis of piping system with NUPIPE computer program, and preparation of stress analysis reports and evaluation of results. Additional assignments included a position with the project engineering group. Duties involved the estimation of man-hours, review and preparation of project, manuals and guidelines, as well as the performance of various project coordination and administrative tasks.

Project Experiences:

Have participated in the design and construction efforts of the following projects:

(1) Lake Michigan Electrical Authority—Watts Bar Units 1 & 2

(2) Illinois Utilities Generating Company—Aurora Electric Units 1 & 2

(3) Carter Atomic Power Laboratory—Carter Facilities Modification Program

(4) Tularosa Electrical Facility—Green Hill Units 1 & 2

(5) Nelco—Peoria Units I & 2

(6) Springfield Community Power Corporation—Springfield

(7) Philippine Electric Company—Manila

(8) Pontiac Electric Co.—Pontiac, Illinois

Skilled in the use of: Lotus 1-2-3, Excel, Word, WordPerfect.

Work Experience:

1996–Present Crown Delivery Service: Courier/Driver (named Driver of the Month three times)

1981–1996 Ethen Services Corporation, Engineer

1978–1981 Ellison and Taper, Associate Engineer

Education:

B.S. in Mechanical Engineering—1978, Chicago Technical Institute. A.A.S. in Applied Science in Mechanical Engineering Technology—1974, Chicago Community College.

References will be provided upon request. Security Clearance: Confidential

MIS MANAGER

A wealth of material is presented effectively and directly. Note the subtle but focused emphasis on recent, relevant positions—and the relegation of others to the brief Prior Positions section.

JOHN SMITH
45 Evansdale Drive, Anytown, STATE/(555)555-5555/E-mail: smith@network.com

GEOGRAPHICAL REQUIREMENTS: OPEN for relocation.

POSITION OBJECTIVE: Director of MIS, Industrial Engineering, Manufacturing

GENERAL SUMMARY: BS Industrial Engineering, MBA Management, over seven years of experience with emphasis on MANUFACTURING INFORMATION SYSTEM, BUDGET and PLANNING SYSTEMS, Government Contract Standards, Quality Control, Task Management, and INDUSTRIAL ENGINEERING.

BUSINESS EXPERIENCE: June 1990 to Present: Stone & Webster Designs
Corporate Headquarters, San Francisco, CA
Program Manager (5/95–Present) Primarily responsible for complete PROGRAM MANAGEMENT and ADMINISTRATION of a company-wide implementation of a COST/SCHEDULE CONTROL SYSTEM which complies with GOVERNMENT CONTRACT STANDARDS, specifically DOD Instruction 7002.1. This System supports $1 Billion in contracts for McDonald-Douglas Airplanes at this MAJOR DEFENSE MANUFACTURER. Activities center on organizing the company-wide implementation teams; serving as LIAISON with GOVERNMENT AUDITORS to demonstrate system for formal validation; acting- as an Internal Consultant to the GM, and developing LONG-RANGE PLANNING. Supervised 37 People.

Design Division, Portland, Oregon
Chief of Manufacturing. Systems (3/93–5/95) Major activities center on the MODIFICATION and IMPLEMENTATION MANAGEMENT of a large REAL-TIME MANUFACTURING INFORMATION SYSTEM on a Honeywell 202 XY System. This system has 100 remote CRT Displays, 500 million records, and is based on the IBM Software Package "IMPACT", a REAL-TIME MATERIAL CONTROL SYSTEM. Duties include upgrading and expanding the system through the implementation of both Radiological and Quality Control Systems, as well as the Project Management of 25 other projects. Supervised 28 Programmer/Analysts. Senior Industrial Engineer (5/92–3/93) Major activities centered on TASK MANAGEMENT of various INDUSTRIAL ENGINEERING Projects and studies related to Division's overall MANUFACTURING OPERATIONS. Specific projects included OPERATIONS ANALYSIS; Material Flow, Procedures and Methods Analysis; Production Breakdown Structures; integration of Computerized Manufacturing Systems, as well as Business Planning; Operational Auditing; and designing a model COD 7002.2 COST/SCHEDULE CONTROL SYSTEM to comply with all Government Contract Standards. Prior positions (6/89–5/92) Included two years as a Financial Systems Analyst in the MIS Division and one year as a QUALITY CONTROL ENGINEER in the Nuclear Division.

EDUCATION
MBA Management, California State University	9/90–5/92
MS Industrial Engineering, Oregon State University	9/86–5/89

The first section places the emphasis on the applicant's strong project development skills.

MARY SMITH
45 Evansdale Drive, Anytown, STATE/(555)555-5555/E-mail: smith@network.com

QUALIFICATIONS
- *Solid background in project development gained during nine years in data processing including: analysis, specifications, design, documentation, programming, testing, training and troubleshooting.*
- *Capable of working with users to obtain an excellent grasp of the problems they want systems to solve.*
- *Ability to relate large numbers of details and exceptions into an integrated design.*
- *Ability to grasp new computer systems and languages very quickly.*
- *Able to develop excellent user and technical documentation to make the system more useful.*

AREAS OF EXPERIENCE
Languages: MVS/ESA, Mvs/xA, Raps, Zeke, Zebb, JES2, RDMS, Omegamon (MVS), Omegamon (ClOS), ClCS, MVSTSO, TLMS, Ca-dispatch, VTAM, COBOL, FORTRAN, PASCAL, C.

EDUCATION
Bachelor of Science: Systems Analysis, 1983
SAN FRANCISCO STATE UNIVERSITY, San Francisco, CA

EMPLOYMENT HISTORY
DATACOM, San Jose, CA   1/92–Present
Shift Operations Supervisor. Currently working as liaison between Operations and Programming departments. This involves conferring with Operations Analyst to check for E & P's; verifying status of system; checking modems; performing IML on central processing unit and/or the 3274 controllers. Generating daily reports; maintaining IR's used for month's end statistics; monitoring and overseeing conversions; upgrading current status of system; assigning codes.
This is a systems software processing company that deals with four major banks and over one hundred branches with an established outstanding record for leaving clients extremely satisfied with their system. Supervisory responsibilities include: Overseeing the activities of a lead operator, Operations Analyst, two tape librarians, and two production control clerks.

BUILT RIGHT CONSTRUCTION, Fresno, CA   6/89–12/91
Assistant Programming Analyst. Designed and implemented three separate systems for purposes of handling payroll reports, temporary employees, and job estimating information.

CALIFORNIA STATE DEPARTMENT OF ENTRY, Long Beach, CA   6/88–9/88
Assistant Programmer. Programmed and evaluated information. This included personnel data, payroll, salary adjustments, budget codes, job recruitment, and health benefits.

ALLIANCE TOTAL COVERAGE INSURANCE, Los Angeles, CA   4/87–6/87
Assistant Programmer. (Internship) Redesigned a data base system for employees; promotion date, skills, and ratings.

LOS ANGELES COMMUNITY COLLEGE, Los Angeles, CA   9/86–3/87
Teaching Fellow Supervised and trained RSTS/S operators. Also responsible for error handling procedures (troubleshooting), user usage and system backups.

References Available upon Request

A summary of relevant accomplishments within a variety of jobs, combined with an intelligently crafted Overview and Education sections, make this resume a standout.

MARY SMITH
45 Evansdale Drive, Anytown, STATE/(555)555-5555/E-mail: smith@network.com

Overview

Eight years of progressive responsibility in system maintenance and upgrade that will allow me to benefit VWI Corporation immediately as a PC System Maintenance Specialist.

EDUCATION

Harding College, San Francisco, CA - presently attending, working toward my B.S. in Computer Science, La Honda Community College - Degree: A.A.S.

EXPERIENCE

9/93–present   *MATERIALS TRANSPORT, INC.*, Fresno, CA
Computer Maintenance. Operations Specialist Duties included working in an I/0 area, monitoring eight PCs and several IBM and 3M printing and microfiche machines. Oversaw system-wide memory upgrade project. Set up payroll, private lines, accounting and production systems—selecting, installing and using a variety of popular software packages.

8/91–8/93   *FEINGOLD TALKING BOOKS*, San Francisco, CA
Tape Librarian. Responsible for answering phones, logging tapes for various users, assisting operators when needed, reinitializing cartridges and recreating new library systems. Created new database system for tracking tape use; developed manual for same.

3/88–8/91   *MERIDIEN TELECOM SYSTEMS*, Fresno, CA
Computer Operator/Data Entry Clerk. Responsible for updating all computer ledgers for customer services, including private lines, accounting, production and payrolls.

6/87–3/88   *COMPUTER SERVICE TECH*, Fresno, CA
Computer Operator Trainee. Duties included maintaining all printers in I/O area, operating the 6670 and 1200 copiers, maintaining the 3800 laser printer, the Versatex machines, and completely supervising an I/O area (section 32).

1/85–6/87   *MATERIALS TRANSPORT, INC.*, Fresno, CA
Receptionist. Duties included filing, typing, fielding telephone inquiries, and managing phone traffic. Assisted in evaluation of new computer system (1986). Developed customized training manuals; performed periodic software upgrades; arranged for routine servicing.

References available upon request.

The first section places the emphasis on the applicant's strong project development skills.

MARY SMITH
45 Evansdale Drive, Anytown, STATE, (555)555-5555, E-mail: smith@network.com

OBJECTIVE
A position in research and development, preferably in the area of applied polymer science

EDUCATION
Ph.D.: Chemical Engineering (expected May 1988) Stanford University GPA 3.8/4.0 Thesis Advisor: Prof. Charles H. Preston
Relevant Coursework: Polymer Rheology; Elasticity Theory; Physical Chemistry of Polymer Solutions; Rational Thermodynamics

M.S.: Chemical Engineering, January 1987 Stanford University GPA 3.5/4.0 Thesis Advisor: Prof. Charles H. Preston
Relevant Coursework: Polymers: Structure and Properties; Synthesis of Macromolecules

B.S.: Chemical Engineering, May 1983 Stanford University GPA 3.4/4.0 Society of American Military Engineers Scholarship

EXPERIENCE
Stanford University, Palo Alto, CA

GRADUATE RESEARCH (Sept. 1983–present) Developed a new theoretical framework for representing polymer-penetrant diffusion problems where significant swelling occurs; designed, constructed, and operated a McBain vapor sorption balance synthesized monodisperse polystyrene latices via emulsion polymerization; determined polymer molecular weights by viscometry; measured crystallinities by DSC and density

TEACHING ASSISTANT (Sept. 1983–present) Lecture and hold recitation sessions in Polymers: Structure and Properties Chemical Engineering Laboratory, Physical Chemistry of Polymer Solutions; evaluate and grade lab reports, exams, and homework

LABORATORY TECHNICIAN (May 1982–June 1983) Designed, constructed, operated, and evaluated membrane processes in the treatment of municipal sewage effluents

MEMBERSHIPS
American Institute of Chemical Engineers, American Chemical Society, Sigma Xi

SKILLS
Scientific programming in FORTRAN and BASIC on various micro/mini/mainframe computer systems, Digital data acquisition using microcomputers, Operation of Differential Scanning Calorimeter, Infrared Spectrometer, Basic electronics and plumbing

INTERESTS
Chinese cooking; personal computers; softball; automobile maintenance; table tennis

Note the concise summaries of success stories comprehensible even to readers without relevant technical backgrounds. (These people may well be screening your resume!)

MARY SMITH
45 Evansdale Drive
Anytown, STATE
(555)555-5555
E-mail: smith@network.com

PROFESSIONAL DEVELOPMENT
Magna Business Services International, Hartford, Connecticut

Senior Associate/Procedures Analyst   **(1990–1997)**
Coordinated teams from Production and Research and Development to eliminate product deficiencies. Set up and monitored steps for tracking and finding problems. Directly supervised the activities of six Manufacturing employees and three Quality Control Personnel. Interfaced with Senior Level Management from various departments, Product Engineering and Manufacturing Engineering. Reported on worldwide installation failures in France, Sweden, Brazil, and Japan. Supervised classroom training to ensure compliance with company policies. Ran weekly meetings to review returning commodity and field incidents.

Associate/Systems and Procedures Analyst   **(1987–1990)**
Responsible for overseeing vital aspects of quality control operations, i.e., to contribute to the overall success of the department and to ensure proper staffing. More specifically, coordinated the activities of the manufacturing personnel for ten departments (three shifts); supervised and delegated work to twenty quality control employees. Monitored statistical reports and implemented corrective action during the manufacturing stages of mainframe computers. Constantly updated and drafted new procedures in order to improve system-wide processes.

Senior Quality Control Technician   **(1983–1987)**
Activities centered on the supervision of eight Quality Control technicians. Drafted work schedules, performance appraisals, troubleshoot problems. Coordinated audits for the department to facilitate the production of mainframe computers. This involved preliminary design stages where data was gathered and reviewed—included setup procedures. Trained new technicians, conducting hands-on testing before certification process.

Quality Control Technician   **(1981–1983)**
Completed Quality Circle Leader and Facilitator training. Worked as a facilitator and assisted with developing a training package for Quality Circle classes for Management.

EDUCATION
COLLEGE OF APPLIED AERONAUTICAL SCIENCES, Hartford, Connecticut
A.A.S. in Engineering, 1981

REFERENCES
Available upon request.

The applicant's "100% self-financed" education offers powerful testimony to his outlook on the world of work, as does his commitment to ongoing education.

JOHN SMITH

45 Evansdale Drive, Anytown, STATE/(555)555-5555/E-mail: smith@network.com

PROFESSIONAL OBJECTIVE

A challenging position in electrical engineering where my professional and educational background in digital and analog circuit application and development, digital signal processing, and test engineering will be of value and allow for professional growth.

EDUCATIONAL BACKGROUND

1997–Present   *SANTA FE CITY COLLEGE, Santa Fe, New Mexico*

Course work towards Master of Science Degree in Electrical Engineering ... Includes course in Electrical circuit Design.

Member of Eta Kappa Nu Honor Society

1997   *SANTA FE CITY COLLEGE, Santa Fe, New Mexico*
BACHELOR OF SCIENCE DEGREE IN ELECTRICAL ENGINEERING

GPA: 3.45/4.0 major ... 3.32/4.0 overall ... Dean's List ... Graduation with Cum Laude honors.

Education 100% self-financed through part-time and summer work ... Positions included Assistant to Contractor, and Draftsman for Maintenance Department.

Special project involved using microprocessors to implement fast Fourier transform algorithm.

PROFESSIONAL EXPERIENCE

1997–Present   *MISSION INSTRUMENTS, Roswell, New Mexico*
Production Engineer (Feb. 98–Present)
Assistant to Project Engineer (June 97–Feb. 98)

Participate in designing digital and analog circuits for Spike recorders and Current Loop recorders ... Company manufactures test instruments for electrical testing needs.

Responsible for testing and troubleshooting work on digital equipment and microprocessor-based three-phase voltage and current recorder.

Supervise two technicians in testing and troubleshooting PC boards.

Use Pascal language in testing optical relay ... Familiar with and skilled in Fortran and Assembly languages.

Able to work hard and develop working rapport with colleagues during projects ... Willing to devote long hours to research activities in order to deliver completely accurate results.

PROGRAMMER (ACCOUNTING BACKGROUND)

The omission of the date of the applicant's first degree is a minor risk, but one probably worth taking. The tactic is designed to encourage an open-minded review of the applicant's qualifications for the position at hand.

JOHN SMITH
45 Evansdale Drive
Anytown, STATE
(555)555-5555
E-mail: smith@network.com

EDUCATION
University of Colorado Springs Bachelor of Arts - January 1994 Major: Computer Science G.P.A.: Computer Science 3.6   Overall 3.5

Springfield College, Springfield, Illinois Bachelor of Arts Major: International Relations

RELEVANT COURSEWORK
- Data Structures
- Discrete Structures
- Operating Systems
- Principles & Survey of Programming Languages
- Database Systems
- Analysis of Algorithms
- Linear Programming
- Self-Study C

COMPUTER EXPERIENCE
Systems: IBM/370 (VM/CMS), VAX/VMS, UNIX, PC (Windows, MS-DOS) Languages: C, COBOL, Pascal, Lisp, Assembler, Prolog, SQL Applications: DBase IV, Lotus 1-2-3, Microsoft Office

EXPERIENCE
1994   *GUARDIAN TRUST CO., Denver, CO*
Customer Service Representative, Handled inquiries from policyholders of Equitable Life Assurance Society and shareholders of Telecom National Corp.

1983–1991   *FUEL ALLIANCE CORPORATION, San Francisco, CA*
Senior Tax Accountant, Analyzed and reviewed tax accruals. Prepared analysis of variances in U.S. source taxable income between quarterly financial periods. Researched the effect of new and proposed tax legislation on Moxway. Drafted memoranda based on research, and presented findings to management. Superior spreadsheet and data management skills.

ACTIVITIES
Dean's List, Ithaca College Association for Computing Machinery
- Member Assistant debugger, Computer Center
- Debugged logic and syntax errors in Pascal
- Assisted students with problems and inquiries
- Certified Public Accountant

References furnished upon request

The resume clearly illustrates a transition from a broad general background
to a strong specific focus of skills and experience.

MARY SMITH

45 Evansdale Drive, Anytown, STATE, (555)555-5555/E-mail: smith@network.com

OBJECTIVE:

A computer programming position which will utilize my educational background and where I may further develop my abilities in the field.

EDUCATION:

HASTINGS COLLEGE, City University of Chicago Bachelor of Arts, June 1984. Major: Computer Science Minor: Mathematics, Grade Point Average: 3.57

ROME INTERNATIONAL UNIVERSITY, Rome, Italy

MILTON ENGLISH COMMUNITY SCHOOL, Viterbo, Italy

COURSES:

Computers and Programming • Linear Algebra Machine and Assembly Language • Differential Equations Discrete Structures • Numerical Analysis and Linear Computer Architecture • Programming Operating Systems • Operations Research Programming Languages • Statistical Methods Database (Graduate Level) • Probability Compilers (Graduate Level) • Demography

SOFTWARE:

Pascal, Fortran, Lisp, Assembly Language Operating Systems - CPV, Virtual Machine SP

HARDWARE:

Xerox Sigma 6/7, IBM 360/370

WORK EXPERIENCE:

Sept. 1985–Present   **BENTON SERVICE MANAGEMENT, INC.**

Consultant to BankCorp International

Assurance and acceptance testing of programs handling a fixed income trading system. Includes debugging, testing modules, setting up test cases and test data to exercise all functions. System released in building block manner; testing performed simultaneously. IDMS database IBM 370 OS environment

July 1983–Nov. 1985   **CHICAGO CITY SAVINGS BANK, Chicago, Illinois**

Bank Teller

Processing customers' transactions according to banking rules and regulations. Balancing daily credits and debits. Handling large amounts of cash.

June 1983–Sept. 1985   **MARKET FILM PROCESSING, Chicago, Illinois**

Sales and Store Management

Preparing film and videotape processing orders. Advising customers on film and processing choices. Maintaining merchandise level and ordering supplies. Taking weekly and monthly inventory. Controlling cash and making deposits. Intensive customer interaction. Promoted to trainer and key store operator, in charge of five other stores.

Sept. 1981–June 1982   **HASTINGS COLLEGE**

Sept. 1982–June 1983

Department of Germanic and Slavic Languages
Student Aide

Clerical office procedures including filing and mass mailing. Handling heavy telephone traffic. Providing information on course descriptions, schedules and professors' office hours.

EXTRACURRICULAR ACTIVITIES:

Member Association for Computer Machinery Student Program, Debugging, Pascal Tutoring, Dean's List

PERSONAL: Single...Date of Birth: 11/7/62...Excellent Health

This applicant has effectively isolated only relevant work experience from a seemingly un-related career field.

JOHN SMITH
45 Evansdale Drive
Anytown, STATE
(555)555-5555/E-mail: smith@network.com

PROFESSIONAL OBJECTIVES
Computer Programmer position with advancement opportunities.

SUMMARY
Recent graduate with a B.S. in Computer and Information Sciences program with specific expertise in FORTRAN, COBOL, Pascal, C, and BASIC; programming experience on a PDP-11 and a VAX/VMS.

SEATTLE COMMUNITY COLLEGE, Seattle, Washington
B.S. Degree in Computer and Information Sciences, Overall G.P.A.: 3.4/4.0

WASHINGTON STATE UNIVERSITY, Spokane, Washington
B.S. Degree in Elementary Education

1988–1991 ST. MARY'S UNIVERSITY, Tacoma, Washington
(Summers) Related course work - D-Base IV.

SPECIFIC SKILLS
Courses in BASIC I and II, FORTRAN I and II, Pascal, COBOL I and II, C, Assembly Language, and Software Engineering I. General knowledge of UNIX.

Programming: Wrote a program as a course project to produce a water bill for a city on a PDP-11 using BASIC language. Wrote a grading system program for a course project on a VAX/VMS using FOR-TRAN and Pascal languages. Wrote COBOL programs on a PDP-11 and VAX/VMS in course projects.

Communications: Participated in definition requirements, design, validation, and Software team meet-ings as part of a Software Engineering course project.

SOFTWARE
Highly skilled in Lotus 1-2-3, Microsoft Excel, Microsoft Word, WordPerfect, and Microsoft Access. At ease with both Windows and Macintosh platforms.

EMPLOYMENT HISTORY
1970–Present TACOMA ELEMENTARY SCHOOL, Tacoma, Washington
Teacher, Grade 5
Duties include participation in public relations projects for the community; supervision and management of students. Coordination and development of school curriculum. Planning, producing, and implement-ing structured student lesson plans, cooperative lesson plans and activities, and special grade projects. Test preparations and evaluations. Prioritizing specific objectives; organizational skills. Participation in Parent Teacher Conferences, teacher grade level meetings, and school meetings. Reviewing textbooks and making recommendations. Handling impromptu situations effectively.

A concise, direct resume that focuses on relevant learning experiences—and
concludes with a single powerful success story.

JOHN SMITH
45 Evansdale Drive
Anytown, STATE
(555)555-5555/E-mail: smith@network.com

EDUCATIONAL BACKGROUND

SAN FRANCISCO STATE UNIVERSITY, San Francisco, CA Current: Graduate School of Arts
and Sciences Studying for Master of Science Degree in Computer Sciences

PORTLAND STATE UNIVERSITY, Portland, OR 1993: Received Master of Engineering Degree
in Industrial Engineering

OSAKA UNIVERSITY, Osaka, Japan 1989: Bachelor of Science Degree in Industrial Engineering

COURSEWORK

Master's courses include: Industrial Computer Technology, Advanced Engineering Economy Advanced Material Control, Digital Systems and Circuits Digital Simulated Technology, Industrial Information Systems Instrument and Control

COMPUTER SCIENCE BACKGROUND

Languages: FORTRAN PASCAL, BASIC GPSS ASSEMBLER Hardware: VAX, CYBER 6502
Software: CP/M Wylbur Script

SPECIAL PROJECTS

For Portland State University, developed a computer program utilized for statistical analysis in the
Department of Industrial Engineering, that was "as innovative as it was glove-fitting," according to
department head Michael Avon.

To learn more about how we might benefit from working together— contact me at the address
above!

References available upon request

PROGRAMMER (DATA ENTRY BACKGROUND)

The inclusion of superior grades and relevant coursework near the beginning of the resume helps to make the case for the transition to a programming environment.

MARY SMITH
45 Evansdale Drive, Anytown, STATE/(555)555-5555/E-mail: smith@network.com

EDUCATION
Pittsburgh Community College, Pittsburgh, PA, Bachelor of Science, December 1989, Major: Computer Science, GPA: 3.77 on a 4.0 scale
Computer Languages
Other Advanced CSc Courses: Pascal
Data Structures and Algorithm Assembler
Software Engineering FORTRAN
Operating Systems C
Database Systems and Data Processing COBOL
Design and Switching Theory Compiler Construction Computer Networks

TECHNICAL CAPABILITIES
Hardware Mainframes: IBM 4341, IBM 3081KX, Microcomputers: IBM PCs, WANG Software: UNIX, WYLBUR, JCL, DOS, OS/MVS, VM/CMS Spreadsheets, Graphics, Word Processing, Data Base Management (DBMS), Database Language: NOMAD, Windows 95

WORK EXPERIENCE
1994–present
Pittsburgh Board of Health, Pittsburgh, PA
Internship: Data Entry, Data Processing, Designed and implemented database applications, updated and entered data, and generated reports in a wide variety of software environments.

1991–1994
Neda's Flowers, Harrisburg, PA
Part-time Assistant, Assisted the manager in dealing with business transactions. Collected and recorded payments.

1984–1991
Chateau Restaurant, Harrisburg, PA
Assistant, Calculated, kept, and summarized business transactions as an assistant for the owners of the restaurant. Helped co-workers and served customers during weekends and holidays.

ACHIEVEMENT
Golden Key National Honor Society

HONOR
Dean's List

SKILLS
Fluent in written and spoken French.

REFERENCES
Will be furnished upon request

The inclusion of specific programs and reports completed is both visually
arresting and relevant to the issue at hand.

MARY SMITH
45 Evansdale Drive, Anytown, STATE/(555)555-5555/E-mail: smith@network.com

EDUCATION

Data Utilities Institute, Cincinnati, Ohio .

Awarded Certificate in Computer Programming and Operations, November 1991.

Awarded General Certificate of Education, July 1990, by the University of Sussex.

Computer Programming and Operations Course Curriculum at Data Utilities Institute:
- Data Processing Concepts
- Basic Operations II Data Processing
- Keypunch •Basic Operating Concepts
- Computer Math
- Financial Concepts
- Systems, JCL, Procedures
- Basic Operations •Systems Design
- Introduction to Computer Operations
- Financial Management
- Hardware CDC
- Cyber-18 •Accounting Fundamentals
- Practical Experience (8 months) on Systems Design, the Cyber-18 Computer System

FORTRAN Programming Programs Written
- Daily Report of Personal Checking
- Customer Filing Account •Program Averages
- Program Factor
- Program Savings Function
- Program Subroutine

COBOL Programming Programs Written
- Transaction Report
- Student College Grade Report
- Insurance Report
- Check Register
- Check Images
- Input Data Validation

Assembler Programming Programs Written
- Program Table •Buy Stock Transactions •Table Loading •Table Processing •Table Sorting •Table Printing

RPG - Reports Written
- Account Balance Report •Customer Account Report •Payroll Report

WORK EXPERIENCE

1978–1980 Dearcher and Daughters, Akron, Ohio (full and part-time)

Responsible for the smooth work flow of all operations. Duties included: inventory control, rotation of stock, packing
out, daily cash receipts, tabulation of accounts and customer service.

PERSONAL DATA

Born 12/19/63...Ht. 5'3"...Wt. 140lbs...Single...In Excellent Health.

REFERENCES UPON REQUEST

The applicant's opening lines give a glimpse of the person behind
the (significant) technical accomplishments.

JOHN SMITH
45 Evansdale Drive
Anytown, STATE
(555)555-5555/E-mail: smith@network.com

A programming specialist committed to the completion of superior work and the identification of opportunities for ongoing professional development.

EXPERIENCE

7/92–Present   *INTERNATIONAL BANK SERVICE, Miami, FL*
Programmer/Analyst
Responsible for the design and analysis of Mortgage, Revolving Credit Loan and Installment Loan Application Programs. - Heavy coding with PLI and COBOL with concern for program structure, language efficiency and maintainability. - Provided assistance to Project Leader in defining and resolving of application program malfunctions and failures. - Specific accomplishments have been: Designing, coding and testing of Mortgage Trial Balance Systems, Mortgage Report Systems, Installment Loan Update and Check Processing Programs.

11/91–5/92   *MARTINI GROUP, Miami, FL*
As a Consultant to Finance Advance Co. (Brokerage Firm), developed and supported a Brokerage and Order Processing System. Responsibilities included: User contact, developing user specifications, program design, coding, debugging, training and support.
As a Consultant to Maximum Finance, West Palm Beach, FL Participated in a project to reduce the response time of operation on Oracle Data Base. Implemented user exits to minimize the imbedded SQL Statement from Total Order Processing Systems. User exits were written in COBOL and performed the functionality testing in the environment of VAX/VMS.

4/90–11/91   *MIAMI DEPARTMENT OF SANITATION, Miami, FL*
Programmer
Coded, enhanced, tested, and maintained programs for Scheduling written in FORTRAN/77 under VM/CMS on IBM 370. - Modified existing SAS programs for changing variables from different data sets to determine most suitable regression analysis models. Used SAS commands to plot graphs.

REFERENCES

Available upon request.

Focus on the applicant's German training and experience establish
the unique elements of this resume.

MARY SMITH
45 Evansdale Drive
Anytown, STATE
(555)555-5555
E-mail: smith@network.com

POSITION: PROGRAMMER

HARDWARE: 4348, 4341

SOFTWARE: COBOL, OS/JCL, OS/MVS, TSO/SPF, VSAM, PROCS, OS/UTILITIES.

EXPERIENCE: March 1984–December 1986
RESEARCH INSTITUTE OF POWER ENGINEERING, Leipzig, Germany

Position: Engineer/Programmer Research and Design Department

Involved in design, writing, testing and debugging new applications and making enhancements of existing programs for accounts receivable, accounts payable, inventory control and depreciation systems and processing of statistical data and experimental results. Participated in cost price and profitability calculation and customizing various report-programs depending on users' requirements.

Unit and system testing, debugging, implementation and writing documentation.

EDUCATION: 1979–1984
POLYTECHNICAL INSTITUTE, Leipzig, Germany

Engineering and Computer Science Degree

REFERENCES: Available upon request. U.S. Permanent resident with employment authorized.

This candidate has specialized international experience which is well documented here.

John Smith
45 Evansdale Drive
Anytown, STATE
(555)555-5555
E-mail: smith@network.com

Resume Position: PROGRAMMER/ANALYST

Hardware: IBM-370/158, 4348, 4341

Software: VSAM, PROCS, OS/UTILITIES

Experience: **April 1987–present** As a staff consultant I worked at client sites as a COBOL Present Programmer/Analyst on a wide variety of applications. These clients consisted of a large New York Bank, and a Manufacturing Company. The applications included credit card processing systems and inventory control systems. The inventory system included several programs which were using variable internal tables loading from sequential files. The files describe the items in the stock, the venders and manufacturers that sell and produce those items, the list price and components of each product, the customers and their orders. The system is used to calculate and control amount and price of incoming materials, parts, tools, outgoing final production, working expenses, and weekly, monthly and annual gross and net profit. The credit card system was processing the customer file to update accounts and account inquiries, to define different groups of customers and their indications depending on users' requirements, and to create various reports for senior management to determine account strategy of the Bank for the future. The programs used internal sort, loading and searching tables technique, subprograms, and parameters.

Daily solution of production problems debugging, customizing of report programs for users, utilizing indexed tables, VSAM and structured COBOL. Ongoing user interface, troubleshooting and program resolution.

March 1984–December 1986 Research Institute of Power Engineering, Osaka, Japan, Position: Engineer/Programmer

Research and Design Department. Involved in design, writing new applications and making enhancements to existing programs and processing of statistical data and experimental results.

Unit and system testing, debugging, implementation and writing documentation, creation various reports.

Education: **1979–1984** Osaka Technical Institute, Osaka, Japan, Engineering and Computer Science Degree

References: Available upon request. U.S. permanent resident with employment authorized.

As this applicant seeks to move into a management level, the use of his unique abilities, the focus on the end user interface coupled with technical skills help to establish the people skills needed for such a move.

JOHN SMITH
45 Evansdale Drive, Anytown, STATE, (555)555-5555, E-mail: smith@network.com

OBJECTIVE
A position in which my experience and training in Programming Analysis can be utilized to our mutual benefit and lead to increased managerial responsibilities.

EXPERIENCE
5/83–Present Guide Insurance Group, Minneapolis, Minnesota
Programmer Analyst
Responsible for the consolidation of information into a database. Interfacing effectively with the Accounting Department to determine requirements for duplicate invoice identification program and Accounts Payable program: created programs to meet requirements. Created program for Personnel Department to facilitate annual EEO Report.. Wrote programs for HIP/HMO Conversion project. transferring accounts into new coverage system. Coding, debugging and maintaining all programs.

3/82–5/83 Transcontinental Minerals, Inc., Minneapolis, Minnesota
Applications Programmer
Responsible for designing, coding, debugging and maintaining of applications for inventory control, sales summary analysis, stock trafficking, accounts payable/receivable and commissions summary for the Accounting and Auditing Departments. The applications were done using BLIS/COBOL Version 4 and a Data General Nova 3 System.

3/81–12/81 Home Protectors Systems, Northfield, Minnesota
Applications Programmer
Responsible for programming a new on-line management system that continuously updates a project's expenditures. This was being done using extended BASIC on a Data General Nova 3 Minicomputer System.

5/80–3/81 Deluth School of Medicine, Deluth, Minnesota
Clinical Technologist

8/77–4/80 Minneapolis Veterans Administration Medical Center, Minneapolis, Minnesota
Research Technician

HARDWARE
IBM 370/145. 4341. Data General Nova 3

SOFTWARE
COBOL. BASIC. PL/I, DOS/VSE JCL

EDUCATION
Harold P. Parrish College of the City of Minneapolis, Currently studying for MASTER OF SCIENCE DEGREE in Computer Methodology.

Courses include:
Operating Systems (IBM/OS), OS/JCL System Design and Analysis, IBM Microcomputers COBOL Programming, Database Input/Output Devices

Founders College of the City of Deluth, Minnesota, 6/77: BACHELOR OF ARTS DEGREE in Biological Sciences

PROFESSIONAL PROFILE
I am a responsible, intelligent individual, interested in further training in and hands-on experience in systems analysis.Possess excellent written and oral communication skills...Capable of establishing an effective working rapport with people of varied backgrounds...I feel the consistent, efficient manner in which I approach all activities and accomplish all tasks can make me an asset to your organization.

References available upon request

Relevant—but indirect—experience in the desired field is presented in the best possible light. The detailed summary of computer language and software skills aids prospective employers in scanned retrieval of the resume.

JOHN SMITH
45 Evansdale Drive
Anytown, STATE
(555)555-5555/E-mail: smith@network.com

OBJECTIVE
Employment as a computer programmer in the field of computer software with opportunities for professional growth and advancement.

EDUCATION
HARTFORD COMMUNITY COLLEGE, Hartford, CT
Bachelor of Arts - Computer Science (May 1996).

HARDWARE AND SOFTWARE EXPERIENCE
IBM 370, VM/CMS, XEDIT, DEC-10,IBM-PC, DOS, DBASE II, PASCAL, ASSEMBLER, FORTRAN, COBOL, LISP, SNOBOL, BASIC, WORDPERFECT, WORD, EXCEL, LOTUS 123.

PROFESSIONAL AFFILIATIONS
Institute of Electrical and Electronics Engineers Association of Computing Machinery.

PROFESSIONAL EXPERIENCE
1/95 to Present
PROGRAMMING CONSULTANT Computing Center of Hartford Community College. Responsible for helping both students and faculty to design, modify, debug their programs, and for providing technical advice whenever requested.

"Established new macros to meet requirements of English Department; work praised as 'superb' by head of Computing Centers."

5/94 to 8/94
TECHNICAL PROGRAMMER Office of Administration, Bureau of Information, Division of School Building, Waterbury City Board of Education. Converted and consolidated inventory programs of all city universities of Connecticut.

1/93 to 12/93
PROGRAMMING CONSULTANT Computing Center of Peters School of Technology. Helped graduate and undergraduate students to debug their programs and provide technical advice whenever requested.

PERSONAL DATA
Willing to travel and/or relocate.

REFERENCES
Furnished upon request.

Appropriate technical accomplishments are introduced by a powerful opening line
that lets the reader know more about the applicant's workplace "mission."

JOHN SMITH
45 Evansdale Drive
Anytown, STATE
(555)555-5555
E-mail: smith@network.com

Committed to delivering flawless technical solutions that make information easier to manage.

TECHNICAL HARDWARE:
IBM 370/3090 Model 200, 4341, under OS/MVS.

SUMMARY SOFTWARE:
OS/JCL, COBOL, CICS/VS, EDF, BMS, INTERTEST IBM utilities, VSAM, TSO/ISPF, VM/CMS, Micro Focus COBOL/2 Workbench.

WORK EXPERIENCE:
1/94–Present INFORMATION PROCESSING TECHNOLOGY, Boston, MA
Programmer/Analyst

Designed, coded and implemented on-line and batch functions for a residual commission reporting system. Wrote residual program maintenance, product type maintenance, residual AE purge and benefit reporting to provide interface with AE benefits statement system. Programmed in COBOL and CICS COMMAND LEVEL.

Took part in the development of a retirement plan inquiry system. Developed and implemented batch and on-line modules (contribution summary, distribution summary and beneficiary display) that allowed IRA personnel clerks to interrogate or view information about customer accounts.

Developed batch modules for the account executive performance reporting system. Designed and implemented the monthly purge cycle. Coded, tested and prepared JCL Procedure. Coordinated the production turnover and trained the users. The system was written in COBOL using VSAM/KSDS files.

EDUCATION:
BOSTON COLLEGE, Boston, MA
Bachelor of Science Degree, 1995 Major: Mathematics Minor: Information System Management

WORCESTER COMMUNITY COLLEGE, Worcester, MA
Associate of Applied Science, 1992

REFERENCES:
Available upon request. *Travel Acceptable, as is Eventual Relocation*

This applicant's academic achievements are enhanced by the inclusion of
short bulleted items under the "Achievements include" heading.

MARY SMITH
45 Evansdale Drive
Anytown, STATE
(555)555-5555
E-mail: smith@network.com

To secure a position as a computer programmer with Cerex Systems.

EDUCATION
Hamford College of the City of Seattle, Washington Bachelor of Science in computer science in June
1992 - GPA 3.85

RELEVANT COURSEWORK
BAL, FORTRAN, PASCAL, LISP, JCL, SNOBOL, COBOL, C, UNIX, D-BASE 3, Filepro, Data-Structures, Data-Base, Programming Languages, Operating Systems.

Achievements include: simulation of an operating system using PASCAL • file management programs
using 370 JCL • object code interpreter using BAL • many business systems in operation written in
Filepro

WORK EXPERIENCE
March 1992–Present
Programmer Frontier Electronics, Seattle, Washington
Responsible for the design, writing and implementation of customized small business software systems
using appropriate databases. Integrated database packages and operating systems . Trained and managed
newer and/or less qualified personnel. Recommended appropriate hardware upgrades/specs.

June 1990–March 1992
Aquatics Manager Continental Recreation Corp., Yakima Park, Washington
Directed day-to-day operation of five pools and staff of 12. Scheduled meetings for non-executive personnel . Authorized expenditures and handled requests for supplies Improved efficiency and cost effectiveness of jurisdiction Arranged activities, fund-raisers, and beautification projects to expand club membership Job facilitated financial independence throughout college.

HOBBIES, ACTIVITIES & HONORS
Member of many athletic teams, High School honors program, Porter Crest High School for Boys

Accepted Early Admissions Program, Yeshiva University.

Organized winter vacation tours and trips.

Organized 1991 American Cancer Society Swimathon, raised $8,000+.

References furnished upon request.

PROGRAMMER (TELECOMMUNICATIONS CALL CENTER EMPHASIS)

Crisp writing, measurable increases in operational speed, plus a powerful Summary of Qualifications combine to make this a memorable document.

MARY SMITH
45 Evansdale Drive, Anytown, STATE/(555)555-5555/E-mail: smith@network.com

OBJECTIVE:
A position as a Computer Programmer with a growth oriented company

SUMMARY OF QUALIFICATIONS:
Excellent communication skills; create and present excellent image of an organization and its services. Understand new ideas and technical concepts quickly, converting these into meaningful results. Efficiently utilize both time and resources.

EDUCATION:
Lynnfield Computer Science Institute, Lynnfield, MA Graduated 1994. Diploma- Computer Programming

Successfully completed an intensive 600 hour course that covered I/O, storage device concepts, multi-programming, structure programming sorts, tables, searches, branches, subroutine linkage, and indexing. All languages were covered extensively. Flowcharted, coded, compiled, debugged computer programs.

Languages
- COBOL, RPGII, III Computer
- SYS/34, SYS/36, AS/400 Applications
- Inventory, Payroll, Accounts Receivable, Sales and Commission Reports, Tax and Invoice Report Master File Update

WORK EXPERIENCE:
6/94–Present TELECOM SCIENCES, INC., Lowell, MA

My duties as a Programmer primarily consisted of: heavy coding in RPGII and RPGIII, Debugging, Documentation, Data Validation, and problem analysis on software used by telephone service reps.

Conversion of System/36 to Native As/400 resulting in increased performance capabilities; creating and bringing files to database.

Redesigning and documenting existing interactive programs, so that they run up to 25% faster. Working out of following client offices: KSMA, Lorraine Linens, Richloom Fabrics and Trafalgar Tours.

6/90–6/94 CREST SERVICE COMPANY, Worcester, MA

Duties included the supervision of about 8-10 other taxicab drivers as well as dispatching and bookkeeping. Eventually assumed managerial level responsibilities.

MILITARY SERVICE:
United States Army, Camp Pendleton, California. Held the position as both a team and squad leader with the Infantry Rank. Received honorable discharge on May 1 1990.

ADDITIONAL EDUCATION:
Harriet Tubman High School, Somerville, MA. Graduated June 1985

REFERENCES UPON REQUEST

This resume is brief and to the point. It illustrates well the candidate's proficiency in the very specific area she wishes to pursue.

MARY SMITH

45 Evansdale Drive, Anytown, STATE/(555)555-5555/E-mail: smith@network.com

PROFESSIONAL OBJECTIVE

To secure a challenging and responsible position in a corporate environment, in which initiative, skills, ambition and commitment to excellence will be utilized to their full potential.

Seeking a position which will allow for continued professional growth, offering an environment in which advancement is based on the strength of individual contributions to the realization of organizational goals.

SUMMARY OF QUALIFICATIONS

Background encompasses experience, accomplishments and the capability to:

Utilize and implement for employer benefit, a demonstrated understanding of the terminology, principles, theory and innovations applicable to COMPUTER/ ELECTRICAL ENGINEERING -"hands-on" working experience with Amiga 2000, IBM s/2, AT&T UNIX, Sys 38 Mod 700; know FORTRAN, Basic, RPG III, Pascal, ASSEMBLY language #68000, 8086 & 6502, dbase Ill +, R Base 5000, Symphony, Harvard Graphics, Boeing Graphics, MIDAS Query. Read manuals, blueprints, schematics or journals as required. Follow required technical, industrial or regulatory guidelines. Work cooperatively with a variety of individuals; interface effectively with coworkers and all levels of management.

EXPERIENCE

Summers 1987, 1988 and 1989 Programmer/Analyst ABINGTON SAVINGS BANK, Abington, Massachusetts Research and make recommendations for PC related applications... interface with user departments for planning, development and implementation of software applications... PC hardware planning, acquisition and implementation...

EDUCATIONAL BACKGROUND

TRENTON STATE UNIVERSITY, Trenton, New Jersey
Bachelor of Science degree program in Electrical Engineering, Minor in Computer Science, 1990 conferral, GPA 3.19. Honors: Dean's List; Correspondence Consultant, Eta Kappa Nu—Honor Society in Electrical Engineering

Diploma, 1986 W.E.B. DUBOIS TECHNICAL HIGH SCHOOL, Newark, New Jersey

Honors/Awards: Salutatorian; Governor's Committee Citation and Faculty Awards-Electronics and English; further awards in Mathematics, Electronics, Science and Scholarship.

Candidate shows interest in the user end of the programming process through her descriptions of the varied projects listed in the resume.

MARY SMITH
45 Evansdale Drive
Anytown, STATE
(555)555-5555
E-mail: smith@network.com

CAREER GOALS	Position as a PROGRAMMER ANALYST in an environment which would enable me to work with people and apply my knowledge of software and hardware to specific projects.
SUMMARY	Five years of COBOL application programming. Heavy user interface. Working experience in CICS/VS Command Level and VSAM files.
SOFTWARE:	CICS Command Level, ANS COBOL, VSAM, OS/JCL, OS/MVS, IBM UTILITIES, TSO/SPF
HARDWARE:	IBM 3033, 4341

EXPERIENCE
1/83–Present UNITED AUTOMATED CONSULTING, INC.
POSITION: PROGRAMMER/ANALYST

- Was responsible for the analysis design and implementation of Bonds Inventory Trading System. It displays inventory when a trader enters a key in a select field. All programs run in a CICS/VSAM environment.

- Participated in the analysis design and implementation of The United Funds Processing System. The System processes funds transfer around the world and controls international and domestic funds. It was implemented in a VSAM/CICS environment.

- Took part in design and implementation of the Foreign Exchange System. The system allows for maintenance of FXR and CIF files. The system was implemented in a COBOL/VSAM environment using CICS Command Level.

- Was involved in the implementation and maintenance of The Personnel System. It was designed to maintain Personnel System Master File. The System was implemented in COBOL/VSAM environment.

EDUCATION
Took various computer related in-house courses.

REFERENCES
Available upon request.

This resume is long but details a long and high-powered career
which includes many leadership roles.

JOHN SMITH

45 Evansdale Drive, Anytown, STATE/(555)555-5555/E-mail: smith@network.com

PROFESSIONAL EXPERIENCE

1/93–Present TechniCom, Lancaster, Pennsylvania, Consultant
Diverse duties include running batch jobs to convert IMS files to DB2 table format with emphasis on troubleshooting, maintenance of DB2 batch/on-line programs using APS macros, tuning, and optimizing performance on application programs and production jobs. Interface with operation and application departments in creating production run books and maintenance of production jobs.

9/91–12/92 Merchants Bank, Pittsburgh, Pennsylvania Consultant
ATM Mini Statement project: Responsible for developing an interface between the Host and Tandem sub-system to retrieve the last five financial transactions from multiple Deposit Account Systems via an ISC and LU 3270 environment.

Customer Account System (CAS) - Multiple Referral project: In this project I was responsible for making major enhancement to the existing "CAS" system to allow multiple referral transactions for Retail/Wholesale Deposit Accounts to be maintained via a Host ISC environment.

3/90–8/91 Free Crest Trust Company, Pittsburgh, Pennsylvania, Senior Programmer Analyst
Project Leader responsible for the design and coding of the REPO Dividend Tracking System. Project life cycle involved writing design and program specs, meeting with DBA, coordinating the integration of other department sub-systems with the REPO system, monitoring progress, managing consultants, and employees, reporting to the manager, coding, and testing programs. The REPO system is written in CICS/BATCH vs/COBOL II which accesses and updates DB2 tables.

Provided system development and technical support for the PUTS Mortgage Backed Payment System. Responsibilities included design, coding, testing batch and on-line programs, reporting to Project Leader, provided technical support to subordinates. The PUTS system is written in CICS/BATCH Vs/COBOL II which accesses and updates DB2 tables.

8/85–3/90 Meridane Koster, Harrisburg, Pennsylvania Senior Programmer Analyst
Assisted in the development of an integrated Trades and REPO System for CPU interface with clearing location. Functions and responsibilities are as follows: Designed, coded and tested CICS programs to processed Trades and verification files before they are released to the clearing location, modified CICS programs to add real-time forward and backward paging, coordinated with other departments in the installation of the CPU system to production, created procs and modified existing procs to create the SOD CPU files for the front end system.

Assigned to support the GSI and MBS back-office areas responsible for implementing the migration of a FTS software package into a central system; provided technical support to staff; designed, coded, tested batch programs to crash files downloaded from an IMS database; created files to feed batch programs to produce financial and statistics reports; created jobs to run parallel to production; modified CICS programs to accommodate new functions. Provided support in the installation of the FTS software package to production.

Modified the FTS Security Master function to lock out MBS users from updating GSI security information. This project involved researching the Security Master function, determining the scope of the change, meeting with users, writing proposal documents, modifying and testing CICS vs/COBOL programs to implementation.

Responsible for integrating the REPO Collateral Substitution function into short tasks. This project included researching the FTS REPO Collateral Substitution function, determining the scope of the change, meeting with functional staff, reporting to the manager, modifying and testing CICS vs/COBOL programs to implementation.

12/79–8/85 Merchants Bank, Allentown, Pennsylvania
Programmer Analyst
Responsible for the enhancements and tuning of the IMS DB Accounting Reconciliation Processing System. This application consists of IMS DB Data Entry programs, Customer Service and PC/XT direct customer dial-up inquiry system. Experience on this system involved rewrites of IMS DB inquiry and data entry programs. Also supervised junior staff members and did technical problem solving for these and other applications.

Developer of the Direct Customer Access System which involved design and coding of a host resident security sub-system to support customer on-line access, via PC, to their checking account information. This project involved creating VSAM files for direct customer access from IMS data bases.

Responsible for technical support and programming for a major Trust and investment software product implementation utilizing IMS DB/DC, VSAM, COBOL, a COBOL generator product.

Designed and coded new modules, and modified a variety of uses such as files conversion and migrations of files. Interfacing to existing DB/DC BAL application (Portfolio Management and Security Movement) software package, and installation of the product modules.

EDUCATION
LANCASTER COMMUNITY COLLEGE (AAS - Data Processing).

REFERENCES
Will be furnished upon request.

PROGRAMMER ANALYST (INSURANCE EMPHASIS)

Candidate documents expertise in insurance industry which includes one very extended position and some bilingual experience.

JOHN SMITH
45 Evansdale Drive
Anytown, STATE
(555)555-5555/E-mail: smith@network.com

POSITION DESIRED: Programmer Analyst

SUMMARY OF QUALIFICATIONS: Three years of Cobol experience. Extensive use of IBM OS/JCL, VSAM files, utilities, and Focus Report Writer. Six months' experience with microcomputer and Relational Data Bases-Language: DataFlex. More than two years of insurance experience.

EXPERIENCE:

3/87–Present UNITED INTERNATIONAL GROUP, Dallas, Texas
Programmer for the Actuarial Department. Responsible for maintaining several systems and files for different divisions. Duties also include designing, coding and testing new Report Systems as requested by the users. Hardware: IBM 3090's. Software: MVS/XA-TSO, OS/JCL, all IBM utilities, Doscan, Expediter, Comparex, Fileaid, Docutext. Languages: Cobol and Focus.

9/86–3/87 ODESSA GRAPHICS, Odessa, Texas
Programmer/Consultant. Responsibilities included development and maintenance of customized database systems for small businesses. Gave technical support and guidance to the users. Extensive use of microcomputer with LAN Novel. Language: DataFlex.

9/85–9/86 CATTLEMORE COMMUNITY COLLEGE, Brownfield, Texas
Part-time Cobol Consultant Instructing students in editing, testing, and debugging Cobol programs on different systems: VM/CMS and MVS with OS/JCL and Wylbur.

9/85–9/86 PINEFORD INSURANCE AGENCY, Odessa, Texas
Personal Lines Manager Assistant Familiarity with all aspects of Insurance with particular specialization in Personal Lines.

10/81–11/82 MANILA INSTITUTE OF STATISTICS, Manila, Philippines Cobol Programmer Duties included designing, coding, and testing validity check routines which were part of a major system. Environment IBM 370 and 4341. Batch programming utilizing OS/JCL.

EDUCATION:

2/84–1/86 CATTLEMORE COMMUNITY COLLEGE, Brownfield, Texas, Certificate Mainframe Computer Programming. Courses included the following: Advanced Cobol, VM/CMS, Systems Design, Assembler, CICS/VS with Ans CoboL, VSAM, etc.

MANILA UNIVERSITY, Manila, Philippines. Full four-year government scholarship to study Computer Science, Equivalent BS obtained in 1982. Courses taken that would be useful in programming work included: Cobol, Fortran, PL/I, Basic, Data Structure, Computer Technology, etc.

REFERENCES: Furnished upon request.

PROGRAMMER ANALYST (LIFE INSURANCE BACKGROUND)

This candidate has experience in being responsible for the entire programming process, from interviewing for user desires through troubleshooting the finished product with the same users.

MARY SMITH
45 Evansdale Drive
Anytown, STATE
(555)555-5555/E-mail: smith@network.com

HARDWARE: IBM 4381, 4341, 370/158, 3033, 3380, 3375, 3350, 3330 Disk, IBM 3278 and IBM 3081 CRT, IBM Laser Printer, 1403 Line Printer

SOFTWARE: COBOL, CICS, DOS/VSE, DOS JCL, Dylacor, TSO, VSAM, OS/MVS, JES2, OS JCL, OS UTILITIES, PROCS, SYNCSORT.

EMPLOYMENT HISTORY:
9/86 to Present
NORTH WESTERN LIFE INSURANCE P.O. Box 1015 Portland, Oregon

Title: Programmer Analyst, Responsibilities:
• Develop and create programs for users.
• 80% CICS programs.
• 20% Batch programs.
• Analyze job specifications.
• Modify, code, test, update and set programs into production.
• Primary work done financial/insurance applications.

5/84 to 9/86
CHRISTINE MOON ELECTRONICS, INC.
4812 SouthEast 38th Portland, Oregon

Title: Programmer Responsibilities:
• Created new report program in Accounts Receivable and Salaried Payroll Systems.
• Member of support group assigned to solution of production problems, debugging, troubleshooting, customizing programs to company needs.
• Handled complete tasks from receipt of company service requests, detail specification writing, coding, compile, unit test and move to production.
• Specialist in preparing VSAM Procs and with either step or job restart.
• Real-time methods for rapid program development and testing.
• Strong structure code, modular code, program clarity for readability and ease of maintenance.
• Debugging abilities.

EDUCATION
Prince Institute of Technology - 1983 to 1984, In-House Courses in VSAM, Structured Cobol, Jobsteam Design.

REFERENCES
Furnished upon request.

This candidate has extensive and varied experience in programming and other systems areas. The variety of projects listed and the illustrated depth to which each was pursued is supportive of the candidate's interest in a job with greater creative potential.

MARY SMITH
45 Evansdale Drive
Anytown, STATE
(555)555-5555/E-mail: smith@network.com

OBJECTIVE
A position as a Computer Programmer/Analyst with a progressive firm providing opportunities for technical and professional growth.

EDUCATION
Fairlane College of the City of Pittsburgh, Pittsburgh, Pennsylvania. Degree: Master of Arts in Computer Science - June 1984

Panama City University, Degree: Bachelor of Arts - June 1979 Major: Education

Major Courses: Advanced Operating System, Principles of Compiler Design, Database, Graph Algorithms, Microprocessors and Programmed Logic, Switching Theory, Artificial Intelligence, Computer Networks, Pattern Recognition, Simulation in GPSS

KNOWLEDGE OF COMPUTER LANGUAGES
PASCAL, PL/1, C (on UNIX), FORTRAN, COBOL, LISP, ALGOL, Assembly Language for Xerox Sigma, GPSS, os/370 JCL and Utilities

KNOWLEDGE OF COMPUTER SYSTEMS
IBM 370 under OS/MVS and JES3, XEROX SIGMA 7 under CP/V, UNIX

EXPERIENCE
Spring/1984 Project: Developed program for implementation of compiler for the subset of C language. Language: PL/1 System: VM/CMS

Project: Designed program for microprocessor-based system with capability to: -Read a thermistor via an A/D convertor -List temperature in degrees Celsius on two seven-segment displays for one minute -Repeat process Language: Assembly Language of Intel 8085A

Fa11/1983 Project: Designed program for management of UNIX-type (tree-structure) file system. Language: PASCAL System: VM/CMS

Project: Designed program for implementing link-deficit algorithm to generate a random graph of degree k. Language: C System: UNIX

Project: Developed program with input spooler simulation capability. Language: PASCAL System: VM/CMS

Fa11/1979 -Madre Junior Middle School, Panama City, Panama Summer/1981 -Teacher: Provided Spanish language instruction for teenage students. -Counselor: Acted as counselor for problem children between the ages of 13–15.

QUALITY ENGINEER (MANUFACTURING SUPPORT SPECIALIST)

Note the "hard numbers" demonstrating performance in a critical area.

JOHN SMITH

45 Evansdale Drive, Anytown, STATE/(555)555-5555/E-mail: smith@network.com

PROFESSIONAL EXPERIENCE

8/93–Present CROWN LANGLEY, INC. Springfield, MA

DC Programmable and Switching Electronic Power Supplies Quality Engineer - Manufacturing Support

Responsibilities: Incoming Inspection Supervision, managing five QA Technicians Supplier quality, and all vendor related problems QA testing, documentation of test procedures and methods; Field Return Analysis and customer C.A.R's. Working together with Design Engineers on corrective actions, Quality cost analysis. Reporting quarterly to upper management

Achievements: Reduced re-works and rejects by 11%. Implementation of SPC and quality awareness programs in various areas using Statistical QC Software on IBM PC. Implementation of Electrostatic Discharge program In the Process of automating final test area

6/91–7/93 UNISON COLOR LAB Electronics Division Lynn, MA

Thick Film Hybrid Products Department - Electronic Ignition Manufacturing and QA Test Engineer

Responsibilities: Developing test plans and implementing necessary hardware and software in Automatic Test Equipment (ATE) based on: LSI-11 Microcomputer from DEC. System used: PDP-11 from DEC Software: Macro4 1 Feasibility Studies on bringing new designs into production test. Documentation of all test procedures and methods QC/QA using Automatic Data Collection System and SPC in a complete CAM environment. System used: VAX 11/780 from DEC. Production support and member of the process problem solving and improvement team

Achievements: Developed test software consisting of 37 tests for the 1988 model year ignition system Revised test software to reduce one second from test cycle time to improve productivity. Production Supervisor: (from February 1992 to August 1992) Managed 22 workers on a completely automated production line. Received in-house training in SPC and Laser Theory and Applications

EDUCATION

1/87–12/87 WESTERN STATE UNIVERSITY Worcester, MA
Continuing education towards a Master's Degree GPA: 3.00 Courses taken: Probabilistic Systems & Random Processes for Engineers - Electronic Project Laboratory

9/83–5/86 WESTERN STATE UNIVERSITY Worcester, MA
Bachelor of Science Degree, GPA: 3.21 Major: Electrical Engineering Minor: Mathematics Concentration: Computer Science1/82–8/83

LOWELL COUNTY COMMUNITY COLLEGE Lowell, MA
Completed freshman requirements GPA: 3.56

Honors: Dean's list all but one semester, Member of Eta Kappa Nu

Computer Languages: Fortran and Pascal

ACTIVITIES

Member of American Society for Quality Control, Member of National Society of Professional Engineers

REFERENCES AVAILABLE UPON REQUEST

A stellar endorsement from a former supervisor adds weight and
credibility to a newcomer's list of accomplishments.

JOHN SMITH
45 Evansdale Drive
Anytown, STATE
(555)555-5555/E-mail: smith@network.com

"Among the most dedicated, responsible, and accurate team players it has ever been my pleasure to work with."—Mel Besson, AIRCO Missile Sciences

OBJECTIVE
Seeking Electrical Engineering position utilizing superior education and experience in Communications, R.F. and Analog applications.

EDUCATION
Massachusetts Institute of Technology (Cambridge, MA), Bachelor of Electrical Engineering, GPA = 3.5/4.0, Graduation Date: June 1997

Lowell Junior College (Lowell, MA), Associates in Arts, GPA = 3.1/4.0, Graduation Date: June 1994

EXPERIENCE
Sept. 1996–present
Research Technician III for the M.I.T. Research Institute (GTRI), Cambridge, MA. Responsible for assisting the Engineers in the Electromagnetic Laboratory, Millimeter-Wave Division in the building and testing of Electronic Projects.

June 1996 to Aug. 1996 Reliability Engineer for AIRCO Missile Sciences, Arrays Division, Colorado Springs, CO. Researched and wrote reports concerning failures and solutions to prevent subsequent failures for the Airborne Laser Tracker Program (ALT). Received Top Secret clearance.

SKILLS
Leadership abilities developed from the positions for Interfraternity Council Cancer Committee; and Social Committee and Leukemia Representative at Sigma Nu Fraternity. Computer abilities developed from courses in Assembly Language and Fortran.
Proficient in German.

HONORS
Dean's List for four quarters.

ACTIVITIES AND INTERESTS
Interfraternity Council Cancer Committee. Leukemia Block Party Chairman. Sigma Nu Fraternity: Leukemia Representative, Social Committee, Philosopher, Bowling Coach, Assistant Soccer Coach. Intramural Sports: Soccer, Frisbee, Football, Bowling, Softball.

GEOGRAPHICAL REQUIREMENTS
New England

An experienced professional makes a compelling case for his return to the workforce after four years as an entrepreneur. Note the emphasis on flexibility and the demonstrated ability to play many roles.

JOHN SMITH
45 Evansdale Drive, Anytown, STATE/(555)555-5555/E-mail: smith@network.com

OBJECTIVE: A position where I may utilize my technical knowledge, proven sales ability, and team-focused working style with a progressive organization where there is growth potential.

WORK EXPERIENCE:

1996–1998 DATATRONICS, Sydney, Australia
General Manager and Owner. Sales of communication equipment, electronic and satellite systems, TVRO, general aviation equipment, VHI networks, and freelance consultation work (gross sales $150,000 annually). Company recently sold.

1994–1996 MELDOC ELECTRONICS, Sydney, Australia
Data Operation Manager. Installation and supervision of installation of Data Communication Equipment, Equatorial Satellite Systems (manufactured in California) and telephone electromagnetic switches. Provided key administrative support as needed.

1983–1984 PACIFIC RIM SATELLITE SYSTEMS, Tokyo, Japan
Microwave Engineer. Installation of TVRO Television Systems in both private homes and commercial establishments; One of two engineers responsible for the redistribution of TV signals using terrestrial links.

1982–1983 ACCUTECH, Melbourne, Australia
Jr. Systems Engineer. Installation of UHF, VHF two way Radio Systems in commercial and private enterprises. Electrical engineering work, electrical wiring in residential homes.

EDUCATION:
NEAR SOUTH WALES GRADUATE SCHOOL, Sydney, Australia Postgraduate work (90% proficiency/grades) Courses Included: - Basic Electronic Data Processing - Cobol Programming

SYDNEY UNIVERSITY, Sydney, Australia BSEE Degree awarded in 1992.
Australian Electrical Engineering Board License issued in 1992.

SEMINARS:
ITT International Telex Exchange Operation (1988), Data Communication Using Satellite Dish for Transmission (1993), Asian Industrial and Trade Fair (1997)

PERSONAL PROFILE:
Self-motivated, flexible. Superior communication skills; able to interact effectively with team members at all levels.

REFERENCES: Furnished upon request.

SENIOR MECHANICAL ENGINEER

With advanced training and experience in programming and Mechanical Engineering this candidate is a high tech double threat.

John Smith
45 Evansdale Drive
Anytown, STATE
(555)555-5555/E-mail: smith@network.com

EDUCATION:
MASTER OF MECHANICAL ENGINEERING IN DESIGN, Lexington Technical Institute, Portland, Oregon
B,E. IN MECHANICAL ENGINEERING, The City College of Portland, Oregon

SOCIETIES: American Society of Mechanical Engineers

CERTIFICATES: Advanced Certificate in HVAC Design

EXPERIENCE:
SENIOR MECHANICAL ENGINEER
(2/88 to Present)   *Thomas A. Peters, Inc., Consulting Engineers, New York*
Responsible for planning, designing of HVAC system for industrial, institutional, and commercial facilities. Significant projects include HVAC design for Willamette Ferry Terminal, Portland, Oregon; HVAC concepts and design of Portland International Airport and Dry Dock No. 2 Modernization, Swann Island Naval Shipyard, Portland, Oregon.

PRODUCTION ENGINEER
(9/85 to 2/88)   *Norwood Elevator Co., Vancouver, Washington*
Drafting and design of elevator cabs; Involved in all facets of production; Supervision of operations.

MECHANICAL ENGINEER
(7/82 to 9/85)   *Kobart Instruments, Milwaukie, Oregon*
Aided in preparation of Correlation Charts, Analytical equations and design Testing of lab instruments.

SENIOR PROGRAMMER, DIRECTOR OF PROGRAMMING
(11/78 to 7/83)   *Statten Parker Securities*
Senior Programmer/Analyst Assisted in writing of CICS programs for on-line money funds order entry system. Assisted other programmers maintain original system.
Supervisor OTC/P&S Departments: installed proper balancing and reconciliation procedures, resulting in substantial savings for the company. Interfaced with the Computer Systems Department with the installation of a new automated order entry system. In control of reconciliation of customer trades made through the company's Canadian subsidiary. Completed three months at Amos Programming School in Waltham, MA.

(9/77 to 11/78)   *Bettington United Clearing Corp.*
Manager R & D/Vault. Responsible for three shifts (24 hours), receipt and delivery of securities from member firms. Netted trade orders and transferred them to net settlement accounts.

(6/68 to 9/77)   *Plyton Trust & Co.*
Manager/Securities Clearance (Buying and Stock Loan Departments) Responsibilities included movement of securities via clearing organizations, balancing positions and money daily, borrowing of securities for third market commitments and processing incoming and outgoing buy-ins.

EDUCATION:
MASSACHUSETTS STATE COLLEGE OF FINANCE
Courses included: Margins (Pershing - 85), Introduction to Back Office Operations (68) and others.

11/80 to 2/81—AMOS PROGRAMMING SCHOOL, Full-time Course

1965 to 1968—MANNY'S HEART HIGH SCHOOL

PERSONAL: Married

REFERENCES: Will Be Provided Upon Request.

Combining management experience with high-powered programming experience, this candidate presents a strong background for her desired career move.

MARY SMITH
45 Evansdale Drive
Anytown, STATE
(555)555-5555/E-mail: smith@network.com

OBJECTIVE
A position with Management-level responsibilities, utilizing my computer-related experience.

EXPERIENCE
11/84–Present   *Meridien Commercial Group, Inc., Youngstown, Ohio*
PROGRAMMER/ANALYST/MARKET RESEARCH CONSULTANT
Responsibilities included: Designing, implementing, and troubleshooting computer systems; updating and revising systems. Overseeing systems training and utilization. Coordinating the activities of 20 sales/marketing and clerical employees. Supervising all computer accounting and inventory systems activities. Preparing and generating all statistical marketing reports; determining low sales items and producing ideas for improved results. -Monitoring activities of shipboard personnel on remote end in Hawaii.

6/82–11/84   *Parker Newspaper, Ltd., Columbus, Ohio*
8/83–11/84: GENERAL SERVICE MANAGER
Responsibilities included: Supervising staff of 70 employees, including all aspects of personnel management. Managing Finance, Administrative and Printing Departments. Designing and implementing systems flows and operational details using IBM PC-XT System. Special Projects: Responsible for overseeing all aspects of the start-up of a new printing division; designing systems; choosing contractors and negotiating contracts; overseeing site preparation and installations; staff development; inventory control and purchasing, including equipment and materials. Interviewing, hiring, training and evaluating all employees.

6/82–8/83: ADMINISTRATIVE ASSISTANT

6/81–8/81   *Priparm Computer Associates, Inc., Springfield, Ohio*
PROGRAMMER TRAINEE

6/76–7/80   *Constantine Business Bank, Cosenza, Italy*
7/79–7/80: BANK MANAGER ASSISTANT
Responsibilities included: Preparing journal accounts and daily, monthly, and annual reports for computer entry.

Summers 1976, 1977, 1978
BANK MANAGER ASSISTANT TRAINEE

EDUCATION
Pratt Institute of Technology, Dayton, Ohio 6/82: MASTER OF SCIENCE DEGREE, Computer Science
St. Mary's College, Cosenza, Italy, 6/79: BACHELOR'S DEGREE, Nutrition and Food Science

TRAINING
CICS Command Level Training Program, Columbus, Ohio 6/82: Received Certificate in Integrated Computer Software.
HARDWARE: IBM PC-XT, IBM 370,4341/II and PDP Dec-10
SOFTWARE: ANSI COBOL, CICS, VSAM,, OS/DOS, JCL, ASSEMBLER, PASCAL AND IMS

SENIOR PROJECT ENGINEER (INTERNATIONAL EXPERTISE)

This resume just gives the outline of an extended career heading
power plant projects in many exotic places.

MARY SMITH
45 Evansdale Drive
Anytown, STATE
(555)555-5555/E-mail: smith@network.com

EXPERIENCE:

Feb. 1989–June 1990   **BASKO ELECTRIC PLANTESKOM, LINZ, AUSTRIA**
HIGH VOLTAGE PLANT SECTION, Senior Engineer
Switchgear Updated, restructured and substantially improved the Company's standard specifications for circuit-breakers and the metal-clad switchgear.

Sept. 1981–Jan. 1989   **BASKO ELECTRIC PLANTESKOM, LINZ, AUSTRIA**
ENGINEERING DEPARTMENT, Design (Project) Engineer
Prepared complete inquiry document for Majuba's Power Station medium voltage switchgear, followed by tenders evaluation and technical recommendation. Designed low and medium voltage standard circuits plus switchgear for Granz and Baden Power Station (each having 3600 MW). Technically managed a low voltage switchgear contract (project) for Tutuka P/S. Designed electrical installations and reticulations for several building services projects (e.g. residential township for 4000 dwellings, including medium voltage substation).

May 1970–July 1981   **INTERNATIONAL ENGINEERS, BERLIN, GERMANY**
Senior Design Engineer, (Jan. 1979–July 1981)/Design Engineer, (May 1970–Dec. 1978)
Responsible for design and feasibility study of various building services and industrial plant projects pertaining to the shipyard of Sczecin. The projects contained lighting (indoor, outdoor), power points, reticulation, switchgear, control circuits, substations, instrumentation. Managed a group of five employees.

Aug. 1968–April 1970   **SHIPYARD OF BERLIN, BERLIN, GERMANY**
Electrical Engineer
Tested and commissioned electrical installations, motors and cables on merchant ship's under construction.

EDUCATION:

THE TECHNICAL UNIVERSITY OF BERLIN   **BERLIN, GERMANY**
M. Sc (Eng) in Electrical Engineering, April 1968

SPECIAL COURSES UNDERTAKEN:
• Analysis and Protection of Electrical Power System, GEC Measurements, Eskom, R.S.A. 1987
• Project Management Program, general course, Eskom, R.S.A. 1986
• P.L.C. Programming and Application, Telemecanique R.S.A. 1985
• Lotus 1-2-3, MS-DOS, Eskom 1988

MEMBERSHIP OF PROF. BODIES:
• The Austrian Council for Professional Engineers
• The Austrian Institute of Electrical Engineers

PERSONAL: Interest in politics, classical music, travelling, playing tennis.

REFERENCES AND FURTHER DATA ON REQUEST

This brief resume establishes experience for this candidate in analysis and system design in bookkeeping and medical fields. These are strong growth areas in which she should be able to advance her career.

MARY SMITH

45 Evansdale Drive, Anytown, STATE/(555)555-5555/E-mail: smith@network.com

PROFESSIONAL EXPERIENCE

1982–1986	*Systems Analyst/Lead Programming Analyst*
MAGNACOM PROFESSIONAL SOFTWARE	*Columbus, Ohio*

Extensive involvement with systems development and maintenance of tax processing system for medium to large C.PA. firms. This system included pro forma of previous years' data, a data entry system, an edit and approval system, and software for preparation of all government forms and schedules necessary for individual tax returns. In 1984 the company introduced the only in-house system for preparation of partnership returns as well as all peripheral support software. Designed and tested an Oil & Gas Depletion/Windfall Profit Tax optional module package. Followed through with interacting with customer support personnel to implement yearly enhancements to package. Worked with in-house accountant and programmers. Integral part of systems developments for the partnership system. Responsible for input design, system design, writing, coding and testing of partner allocations, and Schedule K/K1 calculations. Also met with user groups to design enhancements to systems. Supervised two-three person group projects.

1979–1982	*Data Processing Manager*
MANSFIELD COMMUNITY HOSPITAL SYSTEMS	*Mansfield, Ohio*

In charge of Hospital account in Mansfield. Scheduled computer operations and supervised two shifts of computer operators. Responsible for hardware maintenance and performing necessary program maintenance.

Jr. Programmer/Troubleshooter

Worked variety of locations in the Columbus area. Interacted with Data Processing Managers and hospital personnel designing, writing and implementing specialized reports and/or other modifications to existing batch systems. Projects ranged from listings of patients and amounts due to payroll registers, inventory systems. Also worked extensively on the new D.R.G. Billing System.

Computer Operator

Worked at Springfield Hospital as part of two-person Programmer/ Operator team. Responsible for operation of IBM System 3, Model 15D. Duties included maintaining tape backup library and log; maintaining control totals for all systems (payroll, accounts payable and receivable, and inpatient/outpatient billing); checks and all other printed output.

Languages: RPG II, Basic, PL/I.

Hardware: IBM System 34/36, System 3, 32 & 38, Burroughs 1800, Wang VS.

EDUCATION

COLUMBUS COMMUNITY COLLEGE, Columbus, Ohio
Received Bachelor of Arts in History, 1979. Completed diverse and extensive Liberal Arts Coursework.

References available upon request

This resume includes several strong user application experiences and is highlighted by the candidate's redesign of the "prodigy" software.

JOHN SMITH

45 Evansdale Drive, Anytown, STATE, (555)555-5555, E-mail: smith@network.com

OBJECTIVE

Software Engineer Seeking a challenging position in system/application software field.

EDUCATION

PROVIDENCE COLLEGE, Providence, RI

DATA INFO CORP., Providence, RI

Terminal Application Processing System TAPS/CICS Interface.

STATE UNIVERSITY OF RHODE ISLAND, 9/69 to 5/71 27.Credits completed toward M.A. in Guidance and Counseling. B.A. In English Literature, 6/69 TOKYO UNIVERSITY, Tokyo, Japan.

ENVIRONMENT: IBM AS400, IBM Series I, IBM 370, IBM 4300 series, IBM 8100, Prime 750

HARDWARE: Honeywell DPS6, Sigma 6/7, PDPll/34, Apple II, PS2 and RS6000.

OPERATING SYSTEMS: OS/400 RPS, EDX, VM/CMS, CICS/VS, TAPS/CICS/VS, GCOS6, PRIMOS, DPPX/SP, CP-5, DOS, UNIX, 0S2 and AIX.

LANGUAGES: Cobol, Macro Assembly, EDL, Pascal, Fortran, Sigma 7 assembly, Bal, SQL, C and C++.

EXPERIENCE

6/93–Present   **Programmer Analyst**
THE HAVERLY GROUP - ANALYST CORP.
AS400 Cobol designed, developed, analyzed, coded, implemented, maintained, debugged and tested online (WINS) Haverly Insurance Systems. •Designed and documented specifications and wrote testing plans. •Thru modem, diagnostics and resolutions for users' AS400/WINS problems

9/92–2/93   **EDP Control Officer**
BANK OF TOKYO, San Francisco, CA
AS400/HUB Merchants Bank International System Coordinator Supported and trained 130 nationwide PC users. • Designed, developed, coded and tested in-house PC programs for particular banking tasks.
• Developed, implemented and maintained EDP/O&M policy and procedures manual to comply with governmental regulatory standards, and audited the Bank's internal controls.

3/88–9/91   **System Programmer**
COMMUNICATION SERVICES, CORP.
Designed, analyzed, coded, implemented, maintained, debugged and tested a personal interactive on-line software product -Prodigy. • Conversion of Prodigy from IBM Series/1 to PS2/0S2 and RS6000/AIX. •
Documented and wrote test plans for Prodigy.

2/86–9/87   **Programmer/Analyst**
SYSTEMS COORDINATORS CORP. Providence, RI
Designed, analyzed, coded, installed, and tested manufacturing controlling system and shop floor data collection system.

12/83–2/86   **Member of the Professional Staff**
CREATIVE SOFTWARE, INC. (INFORMATICS) Dallas, Texas
Software audit team member for the Hubble Space Telescope, NASA. •Analyzed for adherence to project requirements, system compatibility, and maintainability.
CREATIVE SOFTWARE, INC., (INFORMATICS) TAPS Division Dallas, Texas
                                                               Programmer/Analyst
Standardized TAPS code for portability and converted TAPS on IBM8100/DPPX. •Installed and Quality Assured TAPS on Prime 750, IBM8100, IBM/OS, IBM/DOS and Honeywell GCOS6. •Designed and wrote TAPS functionalities testing plans and procedures for multienvironments.

11/83–12/83   **Programmer**
NATIONAL COMMUNICATION SYSTEMS, PLANNING & FINANCIAL MANAGEMENT DIVISION Boston, Massachusetts
Developed software to interface with IFPS (Interactive Financial Planning System) and ran A.T.&T. graph system.

4/81–9/81   **CRT/Computer Operator and Programmer**
TECHNOCOM CORPORATION, Computer Department Worcester, Massachusetts
Assisted in installation and initial operation PDP 11/34. •Modified and debugged programs and trained CRT operators.

9/79–5/80   **Lanier Word Processor Operator**
MARTIN KASELYN & CO., Certified Public Accountants Boston, Massachusetts

4/77–5/79   **CRT operator/Order Clerk**
FINANCIAL MANAGERS INTERNATIONAL, INC. Boston, Massachusetts

BACKGROUND
U.S. Citizen; Fluent in Japanese.

References will be furnished upon request.

This brief resume documents an advanced and specific knowledge in solving the issues of generic selection in decision making software.

JOHN SMITH
45 Evansdale Drive
Anytown, STATE
(555)555-5555/E-mail: smith@network.com

OBJECTIVE

A position as a software engineer to design and develop software in the field of Data Communication.

EDUCATION

M.A. in Computer Science, June 1989, Pasadena Community College, Pasadena, California

M.S. in Mechanical Engineering, June 1986, California State University, Los Angeles, California

B.S. in Mechanical Engineering, June 1980, Parma University, Parma, Italy

RESEARCH INTEREST

Expert System Master's degree project: An Approach to Generic Selection Using Knowledge-Based skills. Proposed a conceptual approach to generic selection problem, and integrated the methods for solving Multiple Attribute Decision Making Problems and the techniques of Knowledge Base System to obtain qualified and recommended alternatives. A coding program was set up to simulate the processes of selection.

Computer Network: Studied on the network topology, network model, protocol and network design. Interested in "Micro-to-Mainframe Communication." Software: Kermit, PC-BitCom, PC-ProCom

Database: Studied the database design on the Relational model and the CODASYL model and the database implementation on the SQL/DS and the IDMS/DBTG DBMS. Software: dBASE III Plus

WORK EXPERIENCE

9/87–9/88   Computer Operator (for student expenses)

Paterson Realty, Pasadena, California
Used PC spreadsheet and database software to prepare periodic reports.

COMPUTER KNOWLEDGE

Language: Assembler, APT, Fortran, Lisp, Pascal, C.

Operating System: IBM 370, Cyber 175, VAX-11/780, IBM VM/CMS, UNIX PC: MS-DOS, MS-Window

INTERESTS

Sports, music, chess and bridge.

This candidate has less experience but the sharpness of the focus in the communications areas of electrical engineering give the credentials strength.

John Smith
45 Evansdale Drive, Anytown, STATE/(555)555-5555/E-mail: smith@network.com

Professional Objective
A position in the signal processing, computer systems, telecommunication and software analysis areas of electrical engineering with interest in preliminary design, development and testing.

Education
Boston College, Boston, Massachusetts
M.S. Electrical Engineering Expected May 1989

The City College of Lowell, Lowell, Massachusetts
B.S. Electrical Engineering, May 1986, GPA 3.7/4.0

Experience
Management Summer Employment Program, May 1988–Dec. 1988
Boston Telephone
Supervise non-management employees in administering various activities. Provide technical support and specifications to field crews. Test the applications of new equipment. Repair and maintain telephone switching systems.

Lab Instructor, Sept. 1986–Dec. 1987
Spatial and Temporal Signal Processing Center The Massachusetts State University
Set up and operation of AT&T computers with an ethernet compatible data communication network. Set up and operation of PDP-11/23 workstations with SKYMNK-Q array processor and ILS software package for interactive laboratory systems. Developed and implemented Laboratory experiments such as the effect of sampling, digital filter design and implementation, frequency spectrum using FFT in the TI-PC workstations with TMS320 processor. Provided assistance to students in developing and implementing solutions to their laboratory experiments.

Tutor, Sept. 1983–Dec. 1985
Tutorial Center, The City College of Lowell
Provided assistance to students in undergraduate level physics, chemistry and mathematics.

Honors
National Dean's List - The City College of Lowell, 1985–1986

Dean's List - The City College of Lowell, 1982–1985

Member of Tau Beta Pi (The National Engineering Honor Society)

Member of Eta Kappa Nu (The National Electrical Engineering Honor Society)

Skills
Familiarity with Fortran, Basic, Pascal, C, 6502, 8048 and 8088 language, and with VAX/VMS, RSX-11M +, Micro-RSX, C Shell, IBM VM/CMS, IBM 370/MVS, MS-DOS, CPM-80, Apple DOS, Intel ISIS-II, Intel SDK-85 Monitor. Experienced with Raster Technology model ONE/25S Image Processor, IBM PC Frame Grabber, Intel MDS-800 Development System - Bit Slices. In circuit Emulator, TMS320 XDS/22 Emulator and HP1630 A/D Logic Analyzer

Activities
Student member of Institute of Electrical and Electronic Engineers
Member of Spanish Friendship Association

Personal
U.S. Citizen

This candidate presents many "hands-on" kinds of experiences in the area of solar design. She has attended all the right institutes and in this field with strong grass roots appeal his practical experience will be a respected asset.

MARY SMITH
45 Evansdale Drive
Anytown, STATE
(555)555-5555/E-mail: smith@network.com

JOB OBJECTIVE:
To work with a Solar Energy firm designing and installing Solar Energy systems.

WORK EXPERIENCE:

9/80–6/82   *SOLAR INSTALLER, Design Continuum, Portland, Maine*
Installed and assisted in the design of over forty Solar Domestic Hot Water (SDHW) systems including a Federally funded commercial installation. Responsible for maintaining and troubleshooting these systems. Foreman or assistant foreman on all installations. In charge of all Solar related inventory. Developed a SDHW installation checklist. Wrote owners' manual for SDHW customers. Assistant carpenter on all construction projects, including six active Solariums. Worked with concrete, rough framing, glazing, finish work, dry wall, flooring, roofing, insulating, siding and painting. Often responsible for alternative energy showroom and sales. Helped organize and design showroom. Worked at all trade fairs.

2/80–6/80   *GENERAL LABORER, Bangor College Solar Greenhouse, Bangor College, Bangor, Maine*
Built planters for greenhouse, built heat storage units, installed glazing.

9/79–1/80   *GENERAL LABORER, Project Genesis, Portland, Maine*
Worked as a carpenter's assistant, piper and painter during the construction of the largest Solar Greenhouse in New England at the time.

9/78–5/79   *ASSISTANT MANAGER, Whole Earth Market University of Maine, Portland, Maine*
Student owned and staffed Natural Foods store. Responsibilities included cashiering, ordering, bookkeeping, stocking, etc.

8/78–9/79   *VOLUNTEER, Maine Public Interest Research Group, Bangor, Maine*
Organized University of Maine Energy Teach-In. Organized transportation to Alternative Energy events. Canvassed.

2/76–8/76   *SALES and STOCK, Fiesta Sporting Goods, Inc. Bethel, Maine*
Sales and Stock in camping goods department.

EDUCATION:

1/77–7/80   *UNIVERSITY of MAINE, Portland, Maine*
B.A. in Social Ecology. Program included Energy Studies, Environmental Science, Political Science. Graduated Cum Laude.

6/79–8/79   *INSTITUTE FOR SOCIAL ECOLOGY Hastings College, Blaine, New Hampshire*
Intensive summer program in Solar Energy, Ecological Architecture, Social Theory.

7/80   *SOLAR INSTALLERS CERTIFICATION COURSE Maine*
Environmental Sciences Institute, Portland, Maine A 160-hour course covering all aspects of Solar Installing.

3/82   *ENVIRONMENTAL SCIENCE CONFERENCE New Hampshire*
Solar Conference, Westbrook, New Hampshire A four-day conference on SDHW and the state of the solar industry. Included workshops, lectures, discussions, trade displays.

PROFESSIONAL AFFILIATIONS:
New England Solar Energy Association, Western Maine Solar Energy Association, International Solar Energy Association

CERTIFICATION:
SOLAR INSTALLER Maine Environmental Sciences Institute, Portland, Maine

REFERENCES AND TRANSCRIPTS AVAILABLE UPON REQUEST

A strong practical background in structural engineering, extensive technology application, and wide international experience give this candidate a unique and attractive appearance.

MARY SMITH
45 Evansdale Drive, Anytown, STATE, (555)555-5555/E-mail: smith@network.com

OBJECTIVE
Civil/Structural Engineer, Design or Field Engineering.

EDUCATION
UNIVERSITY OF TORONTO, Toronto, Canada
Master of Engineering Science, 1987

BRAMTON COLLEGE OF TECHNOLOGY, Toronto, Canada
Bachelor of Engineering, Civil, 1980

OSAKA INSTITUTE OF TECHNOLOGY, Osaka, Japan
Bachelor of Science in Civil Engineering, 1970

GENERAL BACKGROUND
Twenty years varied experience in civil and structural engineering with the following firms:

Civil/Structural Engineer, Needham Engineering Co., Needham, Massachusetts July 1993–Present

Civil/Structural Engineer, Bellville & Associates Pty. Ltd., Toronto, Canada, 1987–1992

Structural Engineer, Public Works Dept. of Toronto, Toronto, Canada, 1982–1987

Civil/Structural Engineer, James, Pierce and Morris, Toronto, Canada, 1981–1982

Civil/Structural Engineer, James, Pierce and Morris, Toronto, Canada, 1976–1981

Civil/Structural Engineer, Renfrew Associates, Toronto, Canada, 1973–1976

Structural Engineer, Pacific Rim Corporation, Osaka, Japan, 1971–1973

SPECIFIC EXPERIENCE
STRUCTURAL—Design and documentation from conceptual planning to final details of structural frame works in steel and reinforced concrete for building types such as institutional, commercial, and high-rise multiresidential; and for different facilities such as manufacturing complexes, industrial complexes, cold storage facilities and grain handling facilities. Working knowledge of standards and codes such as AISC, AITC, ACI 318-89, AWS, UBC, API and ASCE 7-88.

CIVIL—Machine, vessel, and tank foundations; retaining walls; grading plans, both rough and finished; storm water drainage design; paving and site development; and minor roadworks.

ADMINISTRATION & SUPERVISION—Team leader and supervisor of drafting personnel, field engineer, project administration, contract administration, project coordination, and engaged private consultants for government projects.

COMPUTER LITERACY—Working knowledge of DOS, Autocad training, WordPerfect, Microsoft Works and practical experience using engineering softwares for structural and civil design applications such as STAAD III, CoGo and Hydra.

PROFESSIONAL AFFILIATIONS
Member - American Society of Civil Engineers

LICENSE
Certified Practicing Engineer (Australia)
Accepted by the Professional Licensing Board (NY) to seat in October 1993 for PE examination (both Parts A and B).

A bilingual candidate in an increasingly international business environment presents some added value for a potential employer.

MARY SMITH

45 Evansdale Drive, Anytown, STATE/(555)555-5555/E-mail: smith@network.com

OBJECTIVE

Desire a position involved in the development of systems software analysis, programming and telecommunication systems.

EDUCATION

Sept. 1981 to Jan. 1983, Winston City College of Winston, Ohio
Graduation with an M.S. Degree in Computer Science.
Courses taken:
- Operating Systems
- Real-Time Processing Systems Database Systems
- Advanced Switching Theory Microcomputer Systems
- Theory of Sequential Machine Compiler Contraction
- Formal Languages & Automata Information Structures Pattern Recognition & Adaptive Systems

Software: FORTRAN, COBOL, ASSEMBLY, PASCAL, C-Language, Intel-8080, JCL.

Hardware: IBM-370, OS/VM.

Jan. 1981 to May 1981, Precision Data Processing Center, Manila, Philippines
Received Certificate in Programming Languages.

Sept. 1976 to June 1980, Chinese Culture University, Taipei, Taiwan, R.O.C.
Received the Bachelor Degree of Law. June 1980

EXPERIENCE

June 1980 to May 1981, Corona Trading Company, Manila, Philippines
Assistant to the President: Assisted in selecting and the import/export of microcomputer software and hardware.

Jan. 1979 to Aug. 1979, Manila Attorney Service Center, Manila, Philippines
Assistant Advisor, part-time: Assisted in the consultation with clients and the preparation of legal cases.

ACTIVITIES

Was on college fencing team and college debating team.

PERSONAL DATA

Date of Birth: Nov. 29, 1956 Sex: Female Visa Type: Permanent Resident Hobbies: Sports

REFERENCES

Furnished upon request.

SYSTEM ANALYST (MARKETING)

The candidate illustrates strong marketing and management skills to be used in support of his System Analyst goals.

JOHN SMITH

45 Evansdale Drive, Anytown, STATE/(555)555-5555/E-mail: smith@network.com

OBJECTIVE: Position using my strong educational background in Management Information Systems and Marketing, communication and leadership abilities and a strong desire for achievement.

EDUCATION:

State University of Texas at San Antonio

BS in Business Management, December 1992

Concentrations in Management Information Systems & Marketing MIS, GPA 3.7, Marketing GPA 3.4, Overall GPA 3.25

COMPUTER SKILLS:

Cobol, Basic, Turbo Pascal, SQL, RBase, dbase, Lotus 1-2-3, WordPerfect, MS DOS, Unix, STP, VM/CMS.

LEADERSHIP:

9/91–12/92. Mexican American Student Organization. Participated in putting up yearly events such as annual dinner, fashion show, picnic etc.

9/91–12/92. Service System Club. Advised on the type of events to be held.

9/91–12/91. Community College Club. Represented Off Campus College Community by attending weekly meetings & participating in events that affected my constituency.

9/90–5/91. Christian Student Group. Elected President; Coordinated activities in an effort to increase member base, which resulted in a 75% increase of people attending. Headed weekly meetings and approved all events.

9/89–5/90. Dorm Activities Coordinator. Included fund-raising and intradorm sports events.

EXPERIENCE:

Gemeo Corporation, Sept. 1993–Present

Fashions Assistant Manager. A good knowledge of all fashion departments, including Ladies, Mens, Infants, Boys, Girls, Hosiery, Fashion Accessories, Jewelry, and Home Fashions. Managed Boys, Girls and Infants Departments as if it was my own business. Made weekly employee schedules, merchandising and softline management decisions, hired and fired associates, and looked at sales trends by comparing last year's and current year's sales on a day-to-day basis, etc.

Texas State University, Summer 1992

Data Entry Operator/Academic Advising Office. Entered transfer student credits into the University mainframe computer system. Office Help/Graduate Admissions Office. Helped secretaries with their work.

Laredo Trading, Inc., San Antonio, TX, Summer 1990

Sold and delivered man-made rugs from Kashmir and India.

Washall Laundromat, Dallas, TX, Summer 1989

Assistant Manager; control of business in the absence of owner.

References available upon request

The emphasis on team sports near the end of the resume sends a subtle but important
positive signal about the applicant's ability to function as part of a group.

JOHN SMITH
45 Evansdale Drive
Anytown, STATE
(555)555-5555/E-mail: smith@network.com

EDUCATION
The City College of San Francisco, B.E. in Electrical Engineering, August 1986 Specialization:
Telecommunications

Specialized Courses:
Computer Networks
Computer Systems Architecture Local Area Networks
Computer Engineering Lab
Local Area Networks Lab
Communication Systems
Digital Signal Processing
Digital Electronics

SPECIAL SKILLS
Computer Languages: Fortran, Assembly Knowledge of the Intel 8085 Microprocessor Wordprocessing
(MS Word), Research codebook formation using SPSSx operating system

Foreign Languages: Spanish, French

EMPLOYMENT
Cybermax Computer Center (10/9/96–Present)
Systems Consultant: Consulting work on IBM PC, PC/XT, PC/AT, PC/RT, System 36/PC, AutoCAD
system, Novell Network (Local Area Network).

Freelance work (9/94–6/96)
Use of oscilloscopes, voltmeters, ammeters, digital multimeters, Frequency Spectrum Analyzer, oscillators, function generators; worked on ECL gates, RAMS, TTL gates, various sequential circuits, FET's.

Parlein's Department Store (12/92–8/94)
Palo Alto, California. Stereos, computers, electronics sales. Floor chief in charge of stereo setup, radio
displays and general register work.

PROFESSIONAL AFFILIATIONS: Institute of Electrical and Electronics Engineers

HOBBIES: Basketball and Handball

EXTRACURRICULAR ACTIVITIES: Intramural handball (Champion twice), Intramural Tennis

References available upon request

Candidate documents very extensive experience in a very specific area - bridge design.

JOHN SMITH

45 Evansdale Drive
Anytown, STATE
(555)555-5555/E-mail: smith@network.com

EXPERIENCE:

March 1994–Present   **JOHN J. KILPATRICK, P.C., Chicago, Illinois**
Design Engineer
Responsible for design of Sterling Freeway. Codesigner of Wankelson Boulevard; Full design of 92'+ 83' Long
Continuous Girders (with CONSYS V5.0—LEAP SOFTWARE); Abutment design with seismic force; check design computation of precast prestressed Box Girder; conducted Geometry Analysis and detailed most DWGs for
the above two bridges.

October 1993–February 1994   **CONECO CONSULTING, Springfield, Illinois**
Design Engineer
Responsible for citywide bridges rehabilitation design and review. Continuous Girder Analysis for 149th Place BR
over URR and Gun Hill BR, etc.

October 1992–October 1993   **ROBERT O'DELL ASSOCIATES, P.C., Bloomington, Illinois**
Senior Engineer, Senior Detailer
Codesigner of 4th Ave. Bridge Detailed Bridge Components (Decks, Abuts, Piers, Joints, Bridge Plans and Elevations, and Profiles, etc.) for Bridges at 65th Place; 102nd St.; Park Place and 4th Ave, etc. Conducted Profile and
Geometry Analysis with Hewlett-Packard vertical curve program Quantity estimate for Bridges.

November 1985–October 1992   **GEOTECHNICAL CONSULTANTS, Marlboro, Illinois**
Engineer, Chief Draftsman (Bridge Department)
Bridge design, detailing, capacity rating, and shop drawing review. Bridge inspection and component rehabilitation
design. Drafting and supervising draftspersons; coordinating work among engineers and draftspersons.
Projects involved: Coswell Bridge (over Marlboro Creek) - Constructed 22 spans of steel trusses and 78 spans of
concrete bents. Inspection and Repair Design.

September 1955–December 1981   **FIFTH DESIGN AND RESEARCH INSTITUTE, Frankfurt, Germany**
Structural Engineer, Design Group Leader
About 26 years, experienced in structural engineering work. Projects included: commercial, residential and industrial buildings; tunnels and underground arsenals; also engaged in some special structural works. Supervised up to
30 engineers and draftspersons.

EDUCATION:

STUTTGART INSTITUTE OF TECHNOLOGY, Stuttgart, B.S.C.E., Structural Engineering, July 1955

REFERENCES:

Furnished upon request

A brief resume with a clear strength in database design in diverse settings.

MARY SMITH
45 Evansdale Drive, Anytown, STATE/(555)555-5555/E-mail: smith@network.com

Career Objective: Full-time position as a Computer Programmer or Analyst, preferably in the field of Database Systems Design, Operating Systems Design and Computer Application Analysis

EDUCATION:
THE CITY COLLEGE OF BIRMINGHAM, Birmingham, Alabama
M.S. Degree in Computer Information Systems, September 1988

THE MASON COLLEGE OF BIRMINGHAM, Birmingham, Alabama
Accounting Major for three semesters
Education 100% self-financed through part-time and summer work . . . Positions included Assistant to Contractor and Draftsman for Maintenance Department.

THE SOUTHERN FRANCE AGRICULTURAL UNIVERSITY
B.S. Degree in Plant Protection

Related Courses:
Business Data Processing
Database Management, Management Science
Economic Issues Information Systems (Computer Network) Statistics Operating Command Languages JCL & COBOL Basic Accounting 1, 2

Honors: Dean's list 1986 (Mason College)

Special Skills: Experienced on DBase 3 plus and Lotus 1-2-3. Familiar with mainframe computers such as IBM-370 VM/SP CMS and IBM PC, XT.

Projects: Designed a relational database system that users can store, search and manipulate relations by using lotus 1-2-3 to analyze companies' economic plans.

EXPERIENCE:
May 1988 to present, Secretary in a Medical Doctor's office

February 1988 to May 1988, Birmingham Transit Authority Station Department, Database system design as an intern student.

December 1987 to February 1988, Carter Fiber Paper Company, Assistant Accountant as a co-op job.

June 1984 to December 1987, Restaurant Assistant Manager.

January 1976 to February 1978, Agricultural Science Centre, Avignon, France. As a Technician.

References available upon request

A strong background is presented in communications and networking.

MARY SMITH
45 Evansdale Drive, Anytown, STATE/(555)555-5555/E-mail: smith@network.com

OBJECTIVE
Position in Network Systems with interest in system integration or design engineering.

EDUCATION
Bachelor of Science in Electrical Engineering/Computer Science, December 1991 POLYTECHNIC UNIVERSITY, Brooklyn, New York

SKILLS
COMMUNICATION SYSTEMS: Knowledge of the hardware and software fundamentals of Local Area Networks and the requirements for network communications.

LANGUAGES: Experience in Pascal, PL/1, Basic, Assembly and Machine Language. Extensive knowledge of data structures and algorithms.

CIRCUITS: Extensive knowledge of solid-state devices, electric and electronic circuits. Theoretical knowledge of switching theory and logic design.

SYSTEMS: Basic knowledge of feedback systems, Digital and analog systems. Knowledge of microprocessors, computer architecture and organization.

WORK EXPERIENCE
August 1989 to June 1991   HOLT, INC., Glendale, California
Upgraded and expanded computer network system. Reported directly to the vice president. Initiated and implemented new inventory control Database that resulted in saving the wastage of raw materials. Assisted in day-to-day operations of company. Actively participated in market development program. Increased customers by 60% over period of two years. Undertook active roll in the factory operations. Reduced customer order delivery time from 15 days to 10 days. Developed production flow technique and control system in the factory. Collected 45% of past due accounts receivables.

May 1987 to September 1987   PACIFIC RIM BANK, Los Angeles, California
In charge of processing customers' applications for Money Market, Day-to-Day, CD and IRA Accounts.

January 1986 to April 1986   FARGO BANK, Glendale, California
Investigated lost card reports and assisted customers in obtaining cash advances and lost card replacement.

HOBBY/ACTIVITIES
Designed and built Heavy Duty 3-Way speakers and sold to Disk Jockeys.

REFERENCES
Will be furnished upon request.

This candidate's experience as a flight attendant illustrates an ability to
serve clients as part of the skills presented.

MARY SMITH
45 Evansdale Drive
Anytown, STATE
(555)555-5555/E-mail: smith@network.com

PROFESSIONAL EXPERIENCE

HORIZON TELEPHONE SYSTEMS   1990 TO PRESENT
Systems Design Specialist
Work closely with corporate clients to assess their telecommunications needs. Analyze blueprints and create station cable records for installation technicians. Interface with local representative on behalf of clients. Assign digital port usage. Write database and program all sets and attendant consoles. Program voice mail and automated attendant. Train all end users, including console attendants and ACD agents and managers as well as voice mail users. Also responsible for job profitability; meeting critical deadlines and managing multiple projects.

PACE UNICOM   1988 TO 1990
Systems Design Representative
Wrote database for electronic key systems, Martin and Corton automated attendant. Trained telephone endusers on NEAX 2400, Martin and Corton voice mail. Also trained on Comdex call accounting.

BAXTER FLIGHT CORPORATION   1987 TO 1988
Flight Attendant
International carrier based at JFK New York. Responsible for passenger safety and comfort.

VICTORY AIRLINES   1978 TO 1986
Senior Flight Attendant   1985 TO 1986
Trained and supervised cabin staff; coordinated inflight services; administered international immigration procedures.

Flight Attendant   1978 TO 1985
International travel. Responsible for safety and comfort of passengers.

BRIGHTON LIFE INSURANCE CO.   1977 TO 1978
Contract Analyst
Wrote group health insurance contracts; provided regional sales support.

EDUCATION AND TRAINING
Boston College   1977
Bachelor of Arts Degree   Major: Sociology/Psychology

TELECOM INTERNATIONAL, INC., Meridian SL-1 Feature Administration - XII Meridian SL-1 Basic Automatic Route Selection - BARS, ACD, MAX

Network Automatic Route Selection - NARS

Demonstrated Qualifications In:
Meridian SL-1 Stadata, Meridian SL-1 Hotel/Motel Feature Administration, Meridian SL-1 ACD (A-C2)

PACE UNICOM PLUS, Station Administration, Customer Relations

Martin and Corton PBX

REFERENCES FURNISHED UPON REQUEST

This candidate shows not only systems expertise but extensive supervision experience.

JOHN SMITH
45 Evansdale Drive, Anytown, STATE/(555)555-5555/E-mail: smith@network.com

POSITION OBJECTIVE
Control Systems Engineer, Design Engineer

GENERAL SUMMARY
BSEE, over 4 years Systems Engineering experience with principal emphasis on HARDWARE SYSTEMS DESIGN, ELECTRONIC CONTROL SYSTEMS, Nuclear Power Plants, Technical Supervision, Closed Loop Control Systems, LOGIC DESIGN and DIGITAL CIRCUIT DESIGN Techniques.

BUSINESS EXPERIENCE
February 1974 to Present: Dallas Power Company, Dallas, TX
Engineer Associate - Project Supervisor, Control Systems (12/75–Present)

Primarily responsible for TECHNICAL SUPERVISION of the HARDWARE SYSTEMS DESIGN of certain key ELECTRONIC CONTROL SYSTEMS for two major NUCLEAR POWER PLANTS. These plants, the Bristol and the Thomaston Nuclear Stations, are currently in the initial Design & final Implementation Stages, respectively. Activities center on directing Hardware Design of 42 different Control Systems for both of the two Plants, utilizing both LOGIC DESIGN & DIGITAL CIRCUIT DESIGN, as well as some Analog Circuit Design Techniques. The major functions of these Control Systems include Electronic Protection, Environmental Monitoring, Total Plant Security, and CLOSED LOOP CONTROL of the Nuclear Steam Supply. Principal achievements: successfully developed Bristol's In-House Security System for classifying and handling/controlling all the sensitive Nuclear Documents. Technically direct six Engineer Assistants, and 41 Design Support Personnel.

Engineer Assistant - Control Systems (2/74–12/75)

Principal activities centered on the HARDWARE SYSTEMS DESIGN & ENGINEERING of ELECTRONIC CONTROL SYSTEMS for a two unit (2 million kw.) NUCLEAR POWER PLANT. This involved complete responsibility for Hardware & Instrumentation Design of 12 different Control Systems using LOGIC DESIGN and DIGITAL CIRCUIT DESIGN, as well as several Analog Circuit Design Techniques. Specific Systems included: Plant Security, ENVIRONMENTAL MONITORING and CLOSED LOOP CONTROL of the Nuclear Steam Supply. Duties included developing Predesign Equipment Specs, as well as writing the System Test Procedures. Technically supervised nine Designers.

EDUCATION
BSEE (3.40/4.00), California Institute of Technology, 6/70–2/74

PERSONAL
Age 26, Married, two Children, Height 5'10", Weight 150 lbs., U.S. Citizen, Excellent Health.

REFERENCES
Available upon request.

SYSTEM PROGRAMMER (DATA PROCESSING)

A long experience in databases, data processing, and documentation highlights this resume.

JOHN SMITH
45 Evansdale Drive
Anytown, STATE
(555)555-5555/E-mail: smith@network.com

SUMMARY

Over fourteen years of Data Processing experience, with twelve in Programming. A Project Manager/Leader with full development life cycle background. Technically proficient with on-line and database systems in the industries of RETAIL, BROKERAGE, MANUFACTURING, BANKING and INSURANCE.

HARDWARE/OPERATING SYSTEMS IBM 4381, 4341, 3090, 308x, OS/MVS/XA, OS/VS1, DOS/VSE, VM, IBM PC/AT

SOFTWARE CICS (Command/Macro), VSAM, IMS/DLI, IDMS/DB-DC (Training), CEDF, BMS, IN-TERTEST, EZTEST, ADS, MFAST, COBOL, BAL, BASIC, JCL (OS and DOS), DYLAKOR, SYSM (ELEC MAIL), CA-UCANDU, TSO/ISPF, ROSCOE, ICCF, OLLE, VM/CMS/SP, PANVALET, LI-BRARIAN, IBM (COPICS/COS), CA/SI (AP and GL), MSA (IPS/POP/AP and HRMS).

PROFESSIONAL EXPERIENCE

LINNARD AND PARKER CORPORATION   **1/85–Present Boston, Massachusetts**

Systems Manager. Headed a department of 15 Senior P/A and Project Leaders, who were retained under the Linnard and Parker consolidation plan with responsibility for.

Supporting all existing systems (24-HOUR DISASTERS, NORMAL MAINTENANCE, plus EN-HANCEMENTS/IMPROVEMENTS). Coordinating the efforts from ANALYSIS through INSTALLA-TION of each SABRE COMMON SYSTEM that replaced all or any part of our original applications. Moving the programming department from Hempstead to Brooklyn.

Prior to the takeover, was responsible for a staff of ten (ranging from Project Leaders to Junior Programmers), for DEVELOPMENT and MAINTENANCE projects in the Application areas of: Sku-based systems - Warehouse Operations, UPC, Price-Lookup, Sales Reporting of Fashion and Big Ticket merchandise.

Employee Productivity/Commission system - Sales and Payroll capture, performance and appraisal reporting, variable rate structures, CICS (controls, adjustments, reporting).

Maintenance of CA (SI) Accounts Payable, General Ledger packages.

CICS Activities: Designed, programmed, and implemented a company-wide security system that all in-house applications interfaced with. Constructed and conducted CICS Training classes. Led a project to develop departmental standards for CICS.

ASSOCIATIVE COMPUTING, INC., Greensboro, North Carolina   **9/81–1/85**

Develop expertise in CICS, IMS/DLI, and VSAM by: Interfacing new programs and exits with the IBM COPICS/COS package. Writing edit, update and print ASYNCHRONOUS programs. Enhancing MACRO level programs under CICS 1.1. Writing dozens of VSAM update and reporting programs. PSB generation.

Lead projects by: Managing eleven programmers in a 3,500 program G/L account conversion. Developing a library management system to control differences between SOURCE and LOAD PRODUCTION and CONVERSION libraries.

System testing, verification and specifications.

User interface and user training.

Produce formal written material by: Creating a booklet for all bank personnel on the purpose, scope, impact and responsibilities of an INTEREST and DIVIDEND TAX COMPLIANCE ACT.

Contributing analytical summations to a "WHITE PAPER" for bank executives on Data Security and Contingency Planning.

Writing a proposal for a potential client.

PRITCHARD, FULLEN AND ODELL, INC., Boston, Massachusetts   9/79–9/81
Programmer Analyst Programmed in CICS/BATCH IMS/DLI and VSAM on various projects, such as: Trading system for management, traders, and brokers. Commission account executive performance system. Commodities ad hoc reporting.

HARLEY'S DEPARTMENT STORE, Somerville, Massachusetts   9/75–9/79
Programmer Programmed in COBOL, BAL, ISAM on Accounts Receivable projects, such as: Descriptive Billing - replaced a card-oriented system. Target Marketing - analysis of mail order and catalog sales. Average Daily Balance - a new finance charge calculation method.

EDUCATION

SOMERVILLE COMMUNITY COLLEGE, Somerville, Massachusetts A.A.S., 1975 G.P.A.: 3.7
HARDWARE: IBM 3090 MVS/ESA, 3084, IBM PC

SOFTWARE: CICS(COMMAND LEVEL), INTERTEST, BMS, SDF, GTB, VSAM, DB2/SQL/SPUFI, IMS/DB/BMP, MFS, ROSCOE, TSO/ISPF, DYLAKOR, EXPEDITER, FILE/AID, PANVALET, LIBRARIAN, MVS JCL, CLIST, LU3270 TYPE PROTOCOL, IBM DISPLAY WRITER, HARVARD GRAPHICS, WORDPERFECT 5.1, PUBLISHER EXPRESS 2.0, LOTUS NOTE, PC/WINDOW.

LANGUAGES: COBOL, COBOL II, BAL LIGHT, RPG LIGHT, METALCOBOL LIGHT, IBM REPORT WRITER, APS MACROS.

This candidate's combination of programming in many languages and communication in both English and Japanese is a strong presentation.

JOHN SMITH

45 Evansdale Drive, Anytown, STATE/(555)555-5555/E-mail: smith@network.com

OBJECTIVE: Seeking a challenging position in system programming where my education and experience would be fully utilized and lead to possible advancement opportunities.

EDUCATION:

COSTA MESA COMMUNITY COLLEGE, Costa Mesa, California
MS Computer Science, December 1991 Major G.P.A.: 4.0/4.0

Special Project: (Graduate Studies) Perfected incisive Pascal Compiler working with theoretical concepts: Lexical, syntax, semantic, Intermediate code, optimization, and code generation (Pascal) C-Language Projects: Worked at interprocess communication facilities to write processes communications system on UNIX for purposes of allowing different users access to same database. Skillfully evaluated output of two master schedulers, Round Robin/Shortest Time Remain First, in order to determine most efficient scheduler. Simulation of microprocessor on UNIX, executing assembly languages with predefined clock cycles and its own microprogram, stack, memory and registers. Designed Database on IBM/VM (SQL-Language).

OSAKA UNIVERSITY, Osaka, Japan B.S. Information Science, 1988

WORK EXPERIENCE:

10/91–Present   *CALIFORNIA STATE UNIVERSITY, San Francisco, CA*
Learning Center/Computer Science Tutor and Grader. Worked fourteen hours a week assisting students having problems with coursework.

6/89–7/90   *OSAKA ARMY COMPUTER CENTER, Osaka, Japan*
Personnel Division/Database Analyst. Activities centered on servicing Division Database (4,000 army personnel). This involved retrieval and updating of files; generating reports on an ad hoc daily basis for supervisor; data input into menu-driven environment. Accomplishment: Charted inquiry system in order for novice users to have access to system.

2/87–12/87   *TOKYO UNIVERSITY, Tokyo, Japan*
Computer Center/System Tutor. Familiarized students with software programs. Worked in two labs; HP 3000 Mainframe, and IBM PC's. Troubleshooting problems. Responsible for verifying student ID's; issuing programs and tutorials.

HARDWARE: IBM 4341, AT&T 3B2, HP 3000, Prime, IBM PC.

SOFTWARE: C-Language, UNIX, SQL/DS, Fortran, VM/CMS, REXX, SQL, HP MPE, Lotus 1-2-3, Pascal, DBase Ill, UNIX System Calls, WordPerfect, WordStar.

AFFILIATIONS: Computer Society of Japan

REFERENCES: Furnished upon request.

Born: July 24, 1954, Fluent in French. Married, no children. Willing to travel.

Job experiences include software and hardware problem solving.

MARY SMITH
45 Evansdale Drive, Anytown, STATE/(555)555-5555/E-mail: smith@network.com

CAREER OBJECTIVE
Immediate Goal: Systems Programmer/Analyst
Long Range Goal: Project Leader/Management

EDUCATION
New York Institute of Technology, Old Westbury, New York 11568, Degree: B.S in Computer Science
Major: Computer Science Minor: Mechanical Engineering, G.P.A.: 3.5

RELEVANT COURSES
Systems Analysis, Systems Programming I, Assembly I, II, Computer Architecture, Discrete Structures, Logic Design, PASCAL I, II, FORTRAN, COBOL, BASIC, RSTS/E (Operators Course at Digital Equipment Corp. -operating system on PDP 11/70) Differential Equations, Engineering Drawing, Engineering Mechanics I, II, Materials Science, Strength of Materials, Thermodynamics.

EMPLOYMENT HISTORY
Forretts Bank Greensboro, North Carolina
September 1981–September 1982   *Computer Operator*
Responsible for all production and runs on a daily basis, identification of both Hardware and Software problems, reporting problems to appropriate area (Software or Vendors); monitoring of telecommunications equipment. Hardware: PDP 11/70, Data General Ecllipse, VAX 11/750.

North Carolina Bank of Commerce and Finance, Winston-Salem, North Carolina
October 1979–September 1981   *Computer Operator*
In charge of all computer related operations. Responsible for on-line CHIPS system operations for P&R function, light involvement in bank reconciliations and correspondence with telex customers. Involved with all I/O functions. Basic operations procedure writing, identification of Software and Hardware problems. Hardware: Data General Eclipse, General Automation (16/65, 16/440), WANG 2200VS.

Dillard and Contry Raleigh, North Carolina
September 1978–October 1979   *Computer Operator*
Responsible for all automated accounting and bookkeeping functions, including posting to General Ledger and running payrolls.

HONORS
Award for outstanding work from the Operations Research and Systems Analysis Dept. of the High Points Technical Institute of High Point, North Carolina, for work done while participating in a National Science Foundation summer program (1975)

MILITARY EXPERIENCE
Six years in the Air Force Reserves. Presently applying for a commission in the Navy Reserves.

References available upon request

Candidate documents many years of a technical systems background and
strong recent supervision and management experience.

MARY SMITH
45 Evansdale Drive
Anytown, STATE
(555)555-5555/E-mail: smith@network.com

OBJECTIVE
A management-level position in Technical Sales where my experience, training and knowledge of computer and telecommunications equipment can be utilized to our mutual benefit and lead to advancement.

EXPERIENCE

1/83–Present   *Crestline Information Systems*

6/84–Present
SYSTEMS MANAGER, California Sales Branch, Sacramento, CA
Responsibilities include: Supervision of 15–20 Technical Consultants and all aspects of technical support operation. Overseeing the applications development, design and implementation of hardware and software products. Development of Branch Technical Support Plan and all Systems Assurance procedures. Coordinating technical support activities for entire product line including: Digital PBXs: Systems 75/85. Networking: ISN, ETN, DCS, Data Networking, X.25, T1. Networking Computer Products: PC 6300, PC7300, 3BI, Office Automation, 3B2, 3B5, 3B15, Unix Operating System, Applications Software. Support of major national accounts including Primary Electric and Copy Quick.

1/83–6/84
TECHNICAL SUPPORT MANAGER, Large Systems, New York, NY
Responsibilities included: Technical support for large PBX and networking systems: System 85. Dimension and Electronic Tandem Network (ETN). Coordinating training activities for New York area field sales force (Account Executives, Technical Consultants). Responsible for the support of 50 national accounts. Developing training programs. Introduction of new products. Daily technical support of marketing sales force. Supervision of other technical support managers in related product areas.

9/82–1/83
SPECIAL ASSIGNMENT, Fresno, CA
Assignment to Early Support Program for beta test installations of System 85 Digital PBX. Responsibilities included: Sales, design and installation of first System 85's at major national account locations. Trained by Bell Laboratories and Western Electric in the design and installation of the product.

5/80–9/82   *San Francisco Telecom Systems*
COMMUNICATIONS SYSTEMS REPRESENTATIVE, San Francisco, CA.
Responsibilities included: Design, pricing and installation of voice and data communications systems: working in industrial sector of Marketing Department in support of Account Executives.

5/78–5/80   *Allied Airlines*
San Francisco International Airport, San Francisco, CA.
PASSENGER SERVICE AGENT Responsibilities included: Ticketing and processing of passengers.

Effective use of the ellipsis (...) helps to support a professional summary
that virtually guarantees review of the remainder of the resume.

JOHN SMITH
45 Evansdale Drive
Anytown, STATE
(555)555-5555/E-mail: smith@network.com

PROFESSIONAL SUMMARY
US Air Force Reduction Officer/Battalion level . . . Maintenance and modification of programming tasks . . .Troubleshooting . . . Awards . . . Army Achievement Medal Military Attache for Core Commander . . . Secret Final Security Clearance . . .

MILITARY AND CAREER RELATED TRAINING
Colorado Springs Air Force University (35 credits) June 1992–February 1994
Miami Dade Business College, Computer Maintenance and Service, 1990

EMPLOYMENT EXPERIENCE
July 1993–February 1995
NORELL COMPUTERS, Miami, FL
Technical Support Specialist
Installed and maintained PC's at regional Headquarters. Able to troubleshoot all types of problems. Hardware: IBM, Packard Bell, Zenith, CD Rom, printers, floppy drives.

July 1991–June 1994
U.S. AIR FORCE, Manchester, England
Military Attache/Force Reduction Officer
Activities centered on preparations for weekly staff meetings for the Lieutenant Colonel with Core Commanders, Company Commanders and other higher level management. Handled transportation arrangements. Produced and maintained graphical data; provided technical assistance, hardware/ software, for 50 offices; maintained evaluation reports and gathered personnel data; implemented Dbase program to control and track equipment; controlled budget requisitions, interacting with logistical shop, dealing with financial matters, buying and repairing equipment. Assisted new personnel. Secondary responsibilities centered on disbanding of a Battalion of 2,000, including families (120 data/text fields). Arranged for redeployments, attending to every last detail. Designed and implemented Dbase program for tracking purposes. Supervised five Administrative Assistants.

March 1990–January 1991
PAXTON'S LUMBER COMPANY, West Palm Beach, FL
Customer Service Representative

August 1987–January 1989
UNITED FIBER PROCESSING CO., Tallahassee, FL
Functioned as a Process Tender, as part of manufacture of large volume of paper. Troubleshooting.

COMPUTERS
Proficient in DOS, Windows, WordPerfect 5.0, 5.1, 6.0 for Windows, MS Word for Windows, Excel, Freelance, Harvard Graphics, Dbase III+, IV, Works for Windows, Norton Utilities, Communication programs, Working knowledge of Novell.

References Available Upon Request

The summary of military experience (which includes essential supporting information) is both powerful and persuasive.

JOHN SMITH

45 Evansdale Drive, Anytown, STATE/(555)555-5555/E-mail: smith@network.com

EXPERIENCE

BALTIMORE TELEPHONE COMPANY, Baltimore, Maryland
1989–present   *Manager*
For three years as a 1st level manager - I managed approximately 15-20 technicians' work schedules and promotion/quality evaluations, handled any technician's discrepancy and effectively oversaw customer relations.

BALTIMORE TELEPHONE COMPANY, Baltimore, Maryland
1983–1989   *Technician*
Installed dial line circuit, leased circuit and remote carrier units for government, private, and public telecommunication facilities. Perform Bell system standards initial site survey, cabling, grounding, inside wiring, terminal blocks/jacks and test circuit specification as per subscriber. Use various electrical meters to perform troubleshooting of equipment and loop circuit from central office frame to demarcating point. In addition, manipulate or transfer underground/aerial cable pair to acceptable conditioning for proper transmission. Maintain daily interaction with engineering, construction, and other departments for accurate record to reduce error and troubles.

Project Assignments: Banning State Hospital - Installed intrabuilding cabling and blocks for SLC 96 cut-over and test intellipath installation.

Maryland City Police - Precinct 107 - Installed wiring, block/jacks for intellipath, and tested D.I.D. trunk lines.

UNITED STATES ARMY, Fort Piedmont, Indiana
1977–1978   *Tactical Field Wireman*
Perform instantaneous communication network setup and removal in the field from buried/aerial cabling to the appropriate units switchboard for an artillery battalion in transition, approximately 10-15 battery units for battalion during each field exercise. Entire coordination of the battalion might consist of approximately ten miles radius.

EDUCATION

BALTIMORE TELEPHONE TECHNICAL TRAINING CENTER Installation/Repair Certificate - 1983

CITY COLLEGE OF HAGERSTOWN, B.S. Geography, 1982

UNITED STATES TRAINING CENTER, Tactical Field Wireman Certificate, 1977

Passing reference to summer employment (in the Personal section) makes the
most of past administrative and supervisory work — and highlights the
applicant's ability to respond creatively in an unfamiliar situation.

JOHN SMITH

45 Evansdale Drive, Anytown, STATE/(555)555-5555/E-mail: smith@network.com

PROFESSIONAL OBJECTIVE

A position in the signal processing, computer systems, telecommunication and software analysis areas of electrical engineering with interest in preliminary design, development and testing.

EDUCATION

LaSalle University, Mansfield, Ohio M.S. Electrical Engineering, Expected May 1989 The City College of Akron, Akron, Ohio B.S. Electrical Engineering, May 1986 GPA 3.7/4.0

EXPERIENCE

Telecommunications Manager

May 1998–present

Apex Telephone. Supervise non-management employees in administering various activities. Provide technical support and specifications to field crews. Test the applications of new equipment. Repair and maintain telephone switching systems.

Lab Instructor

Sept. 1996–Dec. 1997

Spatial and Temporal Signal Processing Center, The Ohio State University. Set up and operation of AT&T computers with an ethernet compatible data communication network. Set up and operation of PDP-11/23 workstations with SKYMNK-O array processor and ILS software package for interactive laboratory systems. Developed and implemented Laboratory experiments such as the effect of sampling, digital filter design and implementation, frequency spectrum using FFT in the TI-PC workstations with TMS320 processor. Provided assistance to students in developing and implementing solutions to their laboratory experiments.

Tutor

Sept. 1993–Dec. 1995

Tutorial Center, The City College of Akron. Provided assistance to students in undergraduate level physics, chemistry and mathematics.

HONORS

National Dean's List - The City College of Akron, 1995–1996. Dean's List -The City College of New York 1992–1995. Member of Tau Beta Pi (The National Engineering Honor Society). Member of Eta Kappa Nu (The National Electrical Engineering Honor Society)

SKILLS

Familiarity with Fortran, Basic, Pascal, C, 6502, 8048 and 80188 language, and with VAX/VMS, RSX-11M4 , Micro-RSX, C Shell, IBM VM/CMS, IBM 370/MVS, MS-DOS, CPM-80, Apple DOS, Intel ISIS-II, Intel SDK-85 Monitor. Experienced with Raster Technology model ONE/25S Image Processor, IBM PC Frame Grabber, Intel MDS-800 Development System - Bit Slices. In circuit Emulator, TMS320 XDS/22 Emulator and HP1630 A/D Logic Analyzer

ACTIVITIES

Student member of Institute of Electrical and Electronic Engineers

PERSONAL

U.S. Citizen

A broadly based array of experience is tied together nicely with
a compelling objective statement.

JOHN SMITH
45 Evansdale Drive
Anytown, STATE
(555)555-5555/E-mail: smith@network.com

PROFESSIONAL OBJECTIVE
A position in the telecommunications, signal processing, computer systems, and software analysis areas of electrical engineering where I can use my expertise in preliminary design, development and testing to deliver profitable solutions.

EDUCATION
Macon State University, Macon, Georgia M.S. Electrical Engineering, Degree Expected December 1998
Atlanta City College, Atlanta, Georgia B.S. Electrical Engineering, May 1996 GPA 3.7/4.0

EXPERIENCE
Lab Instructor & Teaching Assistant   Sept. 1996–Dec. 1997
Spatial and Temporal Signal Processing Center The Savannah State University
Setup and operation of AT&T computers with an ethernet compatible data communication network.
Setup and operation of PDP-11/23 workstations with SKYMNK-Q array processor and ILS software package for interactive laboratory systems Developed and implemented Laboratory experiments such as the effect of sampling, digital filter design and implementation, frequency spectrum using FFT in the TI-PC workstations with TMS320 processor.
Provided assistance to students in developing and implementing solutions to their laboratory experiments.
Taught electric circuit lecture for undergraduate students.

Tutor   Sept. 1993–Dec. 1995
Tutorial Center, Savannah City College, Savannah, Georgia
Provided assistance to students in undergraduate level physics, chemistry and mathematics.

Technician   May 1991–Dec. 1992
Preston Sales Corp., Atlanta, Georgia
Quality Control Testing and Repairing of televisions and videocassette recorders

DESIGN & COMPUTER PROJECTS
• Disk Encoding Format Optimization
• Restoration of Images with Signal Dependent Noise
• Signature Analyzer for Digital IC Testing
• Many Microcomputer Design and Interfacing Projects

HONORS
National Dean's List - Savannah City College, Savannah, Georgia; Dean's List - Savannah City College, Savannah, Georgia, 1992-1995; Member of Tau Beta Pi (The National Engineering Honor Society)

SKILLS
Familiarity with Fortran, Basic, Pascal, C, 6502, 8048 and 8088 language, and with VAX/VMS, RSX-11M+, Micro-RSX, C Shell, IBM VM/CMS, IBM 370/MVS, MS-DOS, CPM-80, Apple DOS, Intel ISIS-II, Intel SDK-85 Monitor. Experienced with Raster Technology model ONE/25S Image Processor, IBM PC Frame Grabber, Intel MDS-800 Development System - Bit Slices. In circuit Emulator, TMS320 XDS/22 Emulator and HP1630 A/D Logic Analyzer

ACTIVITIES
Student member of Institute of Electrical and Electronic Engineers

VIDEO TECHNICIAN

Specific, bulleted accomplishments or recognitions support descriptions
of activities in each of the applicant's cited positions.

MARY SMITH
45 Evansdale Drive
Anytown, STATE
(555)555-5555/E-mail: smith@network.com

OBJECTIVE
Seek a position in the field of Electronic Engineering.

EXPERIENCE

2/96–Present   *PIERCE AND KRAMER, Montgomery, Alabama*
Position: Video Engineer

Diagnostics and Repair Department. Repair sound and TV measurement and monitoring equipment.

* Cited for "operational excellence" at President's Award dinner, 1997.

5/95–1/96   *COMMUNICOR NETWORKS, Mobile, Alabama*
Position: Technician III

Engineering Department. Maintained, repaired, and designed television studio equipment, such as Sony color monitors, switchers, VCR's, TBC's, etc.

* Received superior performance evaluations.

2/92–5/95   *FABRITECH CORPORATION, Anniston, Alabama*
Position: Electronics Technician

Test Department Tested, repaired and aligned computer color graphic image recorders, which can photograph TV images on a high-resolution television screen.

* Established new procedure for repairing equipment that saved an average of seven man-hours per job.

1986–1991   *ELECTRONIC TECHNICAL INSTITUTE, Munich, Germany*
Position: Development Engineer

Television Department Developed and designed analog/digital equipment for composite video signal level stabilization and component regeneration.

EDUCATION

MOBILE COMMUNITY COLLEGE, Mobile, Alabama. Advanced Electronics and Microprocessor Courses Completed 1986.

ENGINEERING AND COMMUNICATIONS INSTITUTE, Munich, Germany Graduated 1973 - Degree: B.S.E.E.

PERSONAL

Citizenship: Resident Alien of U.S., Montgomery, Alabama

Appendix
Resume Resources

The following resume firms supplied talent, insights, and/or sample resumes for this book. If you're looking to develop a superior resume for your job search within the high tech industry, and you feel you need one-on-one attention as you craft your resume, these are the people to call.

Executive Resume Career Marketing Services
P.O. Box 79
Cedar Brook, NJ 08018-9998
1-800-563-6359

Resume Center of New York
15-23 120th Street
College Point, NY 11356
718/445-1956
718/445-1296 (fax)

Resumes by James
102-30 Queens Boulevard
Forest Hills, NY 11375
718/896-6856
718/544-3300 (fax)

Index